Influenza

Dr Jeremy Brown trained at University College School of Medicine in London and completed his residency in emergency medicine in Boston. He was the research director in the Department of Emergency Medicine at George Washington University before moving to the National Institutes of Health, where he is now the director of the Office of Emergency Care Research. His opinion pieces have been published in the *New York Times*, the *Washington Post* and *Discover* magazine.

Influenza

The Quest to Cure the Deadliest Disease in History

Dr Jeremy Brown

TEXT PUBLISHING MELBOURNE AUSTRALIA

textpublishing.com.au
textpublishing.co.uk

The Text Publishing Company
Swann House, 22 William Street, Melbourne, Victoria 3000, Australia

The Text Publishing Company (UK) Ltd
130 Wood Street, London EC2V 6DL, United Kingdom

First published in the USA by Touchstone, a registered trademark of Simon & Schuster, Inc., 2018.
This edition published by The Text Publishing Company, 2019.

Cover design by Imogen Stubbs.
Page design by Jill Putorti.

Printed in Australia by Griffin Press, an Accredited ISO AS/NZS 14001:2004 Environmental Management System Printer.

ISBN: 9781911231219 (paperback)
ISBN: 9781925774153 (ebook)

A catalogue record for this book is available from the National Library of Australia.

This book is printed on paper certified against the Forest Stewardship Council® Standards. Griffin Press holds FSC chain-of-custody certification SGS-COC-005088. FSC promotes environmentally responsible, socially beneficial and economically viable management of the world's forests.

Dedicated to the memory of the dead and the living:

Private Roscoe Vaughan of Buffalo, New York, who died of influenza at Camp Jackson, South Carolina, on September 26, 1918. His death helped us better understand the virus that killed him and millions of others.

Autumn Reddinger, whose fight against influenza is a lesson in personal courage and the triumphs of modern medicine.

*To prevent the spread of Spanish Influenza, sneeze, cough
or expectorate into your handkerchief. You are in no danger
if everyone heeds this warning.*

—SIGNS POSTED IN PHILADELPHIA RAILCARS, OCTOBER 1918

There's nothing quite like flu in terms of the risk.

—TOM FRIEDEN, FORMER DIRECTOR OF THE CENTERS FOR
DISEASE CONTROL AND PREVENTION, JANUARY 2017

CONTENTS

PROLOGUE
AUTUMN

Autumn Reddinger was deathly sick. Her lungs were not working. Her heart was so weak that it could no longer pump blood through her body. The only thing keeping her alive was a heart-lung machine. She lay corpse-like in the intensive care unit. Her parents had called their pastor to administer last rites. How would they tell Autumn's young children that their mom, who was raising them alone, was dying from influenza, an illness that was usually shrugged off as a minor inconvenience? That a vibrant young woman who went to the gym twice a week was now, in December 2013, at death's door?

During the Christmas break, Autumn thought she had a cold but soldiered through the holiday with her parents and two young children at her home in western Pennsylvania. Two days later she felt much better and went to dinner with a friend, Joe. When she returned to her house she began texting him, but the messages he received were jumbled and made no sense. She had been totally coherent at dinner, and Joe knew she hadn't had any alcohol. Alarmed, he got in his car and drove to Autumn's home, where he found her confused and weak. He called her parents to watch the kids and drove her to the local hospital in Punxsutawney. She told the emergency room nurses that her lungs were on fire.

The ER physician went through every test: Autumn's lungs sounded clear when he listened to them with a stethoscope. Her pulse and blood pressure were perfect. She had no fever. Her chest X-ray showed no infection in her lungs. Her blood tests were normal, and a rapid test for influenza came back negative. But something wasn't right, so he admitted Autumn to the hospital for observation, just to be safe, and started her on antibiotics.

Autumn's condition quickly worsened. Over the next few hours she became increasingly disoriented and found it harder and harder to breathe. The antibiotics didn't seem to make a difference. The staff made a call to Mercy hospital in Pittsburgh, a two-hour drive away. Autumn's condition was now critical. Transferring her by ambulance was too risky. Mercy sent a medevac helicopter. By the time it arrived, Autumn could no longer breathe on her own. She was sedated, a tube was put into her throat, and she was connected to a ventilator.

Autumn was taken directly to the intensive care unit at Mercy. By now she was coughing up blood, and it became almost impossible to get enough oxygen into her to keep her alive. A chest X-ray showed that her lungs—which had sounded clear and looked entirely normal a few hours earlier—were now filled with pus and fluid. She was given more antibiotics and hooked up to IV medications to keep her blood pressure from dropping. At one a.m. the ICU team called in Dr. Holt Murray, who had trained as an emergency physician and now specialized in critical care. He was Autumn's last hope.

Murray was an ECMO specialist. ECMO, or extracorporeal membrane oxygenation, is the medical term for the technique used by a heart-lung machine. It takes dusky and dark blood, scrubs it of carbon dioxide, fills it with oxygen, and sends it back, red and healthy, into the body. Doctors use it when they perform heart or lung transplants. Since Autumn's lungs were not functioning at all, the machine could step in and do their work.

2

When Murray talks to family members about putting a patient on ECMO, he has very little time—perhaps only a few minutes—to explain the procedure and get their permission. "I don't think we have other options," he will say, though he is careful to be cautious. "ECMO is potentially lifesaving, but it comes with a host of complications."

Families are usually in no state to make an informed decision, and are highly dependent on the doctor to guide them. Autumn's parents, who had arrived at Mercy, agreed with the ECMO plan.

Very quickly Murray inserted a big needle into a vein in Autumn's groin. This would remove her blood and send it to the machine to be scrubbed and filled with oxygen. Another needle went into her neck, where the blood would return. It didn't take long for the ECMO machine to improve her oxygen levels. But then her heart stopped.

Murray and his team of nurses and specialists began continuous chest massage and gave an injection of epinephrine to restart the heart. Brief success. Then another episode. More epinephrine. The heart kicked back into action, but it was barely working. Murray did an ultrasound of Autumn's heart that showed it was functioning at less than 10 percent of capacity. It was no longer able to pump blood through her body.

Doctors have a rather unsavory term for patients in Autumn's condition: "circling the drain." It describes, in cold and colloquial terms, that hopeless feeling when every attempt to save a patient is failing. Autumn was circling the drain.

Even though the initial influenza test had been negative, Murray now repeated it using a much more sensitive technique. It revealed that Autumn had the H1N1 influenza virus, the same virus behind the swine flu outbreak of 2009. In a matter of hours, the virus had destroyed her lungs and was now attacking her heart muscle. The ECMO machine that had taken over for her lungs was no longer good enough. It now needed to take over the work of her failing heart. To do that, the machine needed to be replumbed. And that meant transferring Autumn four blocks to the University of Pittsburgh's Presbyterian hospital, where a cardiothoracic surgeon could do the procedure. Murray traveled with her in the back of an ambulance, carefully monitoring the portable ECMO machine. Autumn was taken directly to the operating room. The surgeon cracked open her sternum with a surgical saw, and then inserted one catheter into her right atrium, one of the four chambers of her heart. A second catheter went directly into her aorta, and her sternum was wired back together. Her chest now had a long, fresh vertical wound and two large tubes sticking out from it, connecting Autumn to the heart-lung machine. This was the end of the road. There were no more machines, no better interventions or heroic measures that Murray could offer. She would either pull through or die.

Autumn's parents, Gary and Bambi, sat with their pastor in a small family room near the ICU. "We got together and prayed," Gary says. "Then the pastor told us that she saw two angels, and that things would be okay."

The pastor was right. Autumn stabilized. Her heart, stunned by the influenza virus, returned to normal over the next few days. Antibiotics cleared up a secondary bacterial pneumonia, and her blood pressure no longer took sudden nosedives. On January 10, 2014, she was disconnected from the ECMO machine, though she was still sedated and attached to a ventilator. A week later she had improved enough to be discharged from the cardiothoracic intensive care unit. After another month of slow improvements, she left Presbyterian hospital on February 13 for a rehabilitation center closer to home. She had beaten influenza, but now had another battle to fight. Patients who have a long stay in the ICU often develop a profound generalized weakness. At the rehab hospital, Autumn had to learn how to walk again, to climb stairs, and to perform a host of other daily activities that she had once taken for granted. After two weeks of rigorous exercise, she left rehab and came home. In the fall of 2014, nine months after she first caught influenza,

Autumn was finally strong enough to return to work. Her medical bills were close to $2 million, but she was fortunate to have good medical insurance and was only on the hook for her $18 copay.

She was left with scars on her neck and chest. As a result of nerve damage from the needles in her groin, she cannot bend her left ankle, and her left leg sometimes goes numb. But her survival and recovery were a triumph of modern medicine. She pulled through because she was close to a medical facility that was able to offer her the best of today's interventions.

Autumn's fate would have been far different during the influenza pandemic of 1918, the worst in recorded history. The best medicine at the time was aspirin, but it was new, misunderstood, and often given in fatal doses. Desperation and ignorance produced plenty of awful methods of "healing," from barbaric bloodletting to treatments with toxic gas. It is estimated that between 50 and 100 million people died during that influenza pandemic. In the United States there were 675,000 deaths, ten times as many as had died in combat in the Great War, which was ending at the same time the flu was peaking.

The flu is something we have all experienced at some time: the winter cough, the fever, the body aches and chills that knock us out for three or four days and then disappear. As an ER doctor and as a patient, I have experienced it from both bed and bedside. The one and only time that I have visited the emergency room as a patient was when I was sick with a particularly nasty bout of the flu. I had a high fever and was delirious. I was too weak to drink or get out of bed, and I was dehydrated. But even the modern medicine that pulled me through my relatively minor infection—and that which brought Autumn back from the brink of death—isn't always enough. The flu is still a serial killer.

We all harbor the desire to see cancer cured and heart disease eliminated. I obviously share these hopes, but as an ER doctor I have found myself wishing for something far more modest: a cure for influenza.

5

We tend to shrug off the flu as just a nasty cold, but in the United States alone it kills between 36,000 and 50,000 people each year. That's an astonishing and depressing number. But here is even worse news: if a pandemic strain of influenza as deadly as the 1918 virus were to infect the United States today, more than 2 million people could die. No conceivable natural disaster would compare. And it's not something we can just pretend won't affect us. In early 2018 newspapers warned that the flu season was "the worst in nearly a decade." All across the country there were reports of young, healthy people dying. Several hospitals were so overwhelmed with the influx of flu patients that they had to set up triage tents or turn patients away.

Flu is certainly not "the emperor of all maladies," as cancer was described by the oncologist Siddhartha Mukherjee, but it is the malady of all empires. It has been with us since the dawn of time, and it has afflicted each civilization and society in every corner of the globe.

We've had several close calls with major viral outbreaks since 1918. The Hong Kong avian flu outbreak in 1997 killed few people, but only because 1.5 million infected chickens were slaughtered before they could spread the disease. In 2003 there was the SARS outbreak that infected at least 8,000 people and killed about 10 percent of them. More recently we have encountered MERS, Middle East respiratory syndrome, which infected 1,400 people between 2012 and 2015. That disease entered the human population courtesy of infected dromedary camels. (Here's some free medical advice: before taking a swig, make sure your camel milk has been pasteurized.) These viral diseases all originated in animal hosts (we think) and somehow jumped into humans—which is what happened in 1918 (we think). We don't know when or where the next viral pandemic will occur, but it *will* occur. There is little doubt that, unless we plan for it, we are in for rather a rough ride.

PROLOGUE: AUTUMN

One hundred years after the pandemic of 1918, we have learned an enormous amount about influenza. We know its genetic code, how it mutates, and how it makes us sick, and yet we still don't have effective ways to fight it. The antiviral medications we have are pretty useless, and the flu vaccine is a poor defense. In good years it is effective only half the time, and in 2018 the record was even worse; the vaccine was only effective in about one-third of those who received it.

Just one century is all that separates us from a global health crisis that killed more people than any other illness in recorded history. What we've learned in the interim is enough to scare and motivate us, but maybe not enough to stop another pandemic from happening. Because of its mystery, and its ability to mutate and spread, the flu is one of mankind's most dangerous foes. The lessons of 1918 may be our only inoculation against a deadly sequel.

1

ENEMAS, BLOODLETTING, AND WHISKEY: TREATING THE FLU

Among my many weaknesses, none is worse than my appetite for chicken soup. When I was a child, I looked forward to my mother serving it on Friday nights. To this day, it brings back memories of growing up in London, and its long and rainy winter nights. For centuries, chicken soup has been seen as a folk remedy for coughs and colds, fevers and chills—all the symptoms of the flu. My mother would remind me to finish my serving so that I wouldn't become ill over the winter. It was the most delicious preventive medicine imaginable.

Many years later, while at medical school in London, I came across a study suggesting that chicken soup might actually be the real thing. The article was published in 1978, in the journal *Chest*, and its title is almost as delicious as the soup itself: "Effects of Drinking Hot Water, Cold Water, and Chicken Soup on Nasal Mucus Velocity and Nasal Airflow Resistance."

In the study, pulmonologists paid healthy volunteers to drink either hot water, cold water, or hot chicken soup, and then they measured any changes in congestion—or, as the title suggests, how fast mucus and air moved through the nasal cavity. The researchers concluded that while hot water is good to help clear your stuffy nose, chicken soup has "an additional substance" that does the job better. No one's clear on what this secret ingredient is, but researchers have theorized that the key to the soup's restorative powers is the nourishing partnership of vegetables and chicken.

Dr. Stephen Rennard of the University of Nebraska Medical Center has studied chicken soup for more than a decade, and in 2000 he found that the recipe from his wife's Lithuanian grandmother reduced upper respiratory cold symptoms by inhibiting the circulation of certain white blood cells that react to infection—meaning that chicken soup is a kind of anti-inflammatory.

"There's little doubt that, one hundred years from now, probably everything else that I've done will be forgotten because it'll be irrelevant and out of date," says Rennard in a YouTube video shot in his kitchen as his wife cooks. "But the chicken soup paper probably will still be cited." Doctor tested, grandma approved.

Sometimes age-old wisdom yields clinical success. I wish I could say the same for other remedies that have been tried for treating the flu. Enemas. Mercury. Tree bark. Bloodletting. Methods to boggle your mind and turn your stomach. Be glad you weren't born in the 1900s. Today, no reputable doctor would prescribe these treatments, but just over one hundred years ago they were state-of-the-art therapies. Perhaps the only thing more shocking than the crudeness of these "cures" is the fact that our state-of-the-art methods in the twenty-first century aren't wildly more advanced.

Not three years after he resigned as the first president of the United States, George Washington was on his deathbed. As a last-ditch effort to save him, doctors opened his veins to thwart the infection ravaging his throat. Washington endured four rounds of bloodletting, the last one only a few hours before he succumbed.

"I am just going," Washington said to his secretary, Tobias Lear, at that point.

"He died by the loss of blood and the want of air," said a family friend and doctor, William Thornton—who suggested that Washington could be reanimated by a transfusion of lamb's blood.

Bloodletting is the practice of draining the body of blood—and therefore, in theory, of toxins and disease—and was mainstream medical practice for more than two thousand years. In the eras before any useful medications or treatments, bloodletting was pretty much all there was. It dates back to at least the fifth century BCE, and is mentioned in the writings of the second-century Greek physician Galen, who taught that it was an important tool that could heal the sick. Bloodletting is frequently mentioned in the Talmud, a work recording the debates about Jewish law and ethics that was finalized around 600 CE, and was widely practiced during the Middle Ages and beyond. One of the most respected medical journals in the world is named after the main tool used in bloodletting: *Lancet.*

Bloodletting never worked. In fact, it was terribly dangerous—just ask George Washington—but it continued to be prescribed for influenza into the first decades of the twentieth century. And not just by fringe practitioners. It was recommended by military doctors who were on the front lines during World War I and saw another enemy—a microbial one—outflanking the ranks. What's more, these doctors wrote about their experiences of bloodletting in important medical journals, including, poetically, the *Lancet.*

Three British doctors serving in northern France in December 1916, about two years before the outbreak of pandemic influenza, described a disease that swept through the army camps with catastrophic results. It was as if the influenza virus were doing a sort of dry run, preparing to unleash even more destruction later. The doctors were certain that the disease, which they called "purulent bronchitis," was caused by the influenza bacillus, and they described their efforts to treat the poor servicemen who were taken ill. They were failing.

"So far," they wrote, "we have been unable to find anything that has any real influence on the course of the disease." And then this: "Venesection has likewise failed to benefit the patient for more than a very short time, though possibly we have not resorted to this treatment sufficiently early."

You could almost miss this if you were reading the paper quickly. But it's there. The British physicians had tried venesection, the medical term for bloodletting, and it had not worked—perhaps, they thought, because they tried it too late in the course of the disease. Two years later, at the peak of the flu pandemic, other British military physicians reported bloodletting their patients, too, only this time they reported that it worked, at least in some cases.

It wasn't only the British who were still bleeding their patients in the twentieth century. In 1915 Heinrich Stern, a physician in New York, published his *Theory and Practice of Bloodletting*. While Stern was critical of bloodletting for most medical conditions, he did believe it could be helpful in some instances.

"I am an advocate of the conditional employment of this ancient method," he wrote, "and I believe it unnecessary to state that I do not consider it a panacea."

Stern was somewhat ambivalent in recommending it as a primary therapy for the flu, but almost a decade later, in America's leading medical journal, doctors were still advocating bloodletting to treat pneumonia, convinced—without a shred of evidence—that it would get results when "our more conservative methods fail."

Bloodletting to cure influenza eventually went out of fashion in the twentieth century, but other wild and suspicious treatments were still part of the medical repertoire.

In 1913 a small book with a black cover was published by a doctor named Arthur Hopkirk. Its title was etched in gold: *Influenza: Its History, Na-*

ture, Cause and Treatment. Hopkirk recommended a series of bizarre treatments for influenza. For fevers, the good doctor recommended "a purge," meaning a laxative, such as the delightfully named "effervescent magnesia." Severe cases of flu required a heavy-duty laxative like calomel, which is made with mercury chloride. Mercury, of course, is highly toxic.

Hopkirk's 1914 recommendations did contain some nuggets of sound advice here and there. For example, alongside the poisonous mercury laxative he recommended aspirin, derived from the bark of the willow tree. (Of course, aspirin is still used, though today you are more likely to take Tylenol or Motrin.) But even that recommendation may have done more harm than good, because doctors didn't yet know how to safely dose the drug. Symptoms of an aspirin overdose start with ringing in the ears, followed by sweating, dehydration, and rapid breathing. A severe overdose results in fluids pouring into the lungs, mimicking the actual symptoms of the flu. Fluids then enter the brain and it swells, resulting in confusion, coma, convulsions, and death. People were not just dying of the flu during the Spanish flu pandemic; they were also dying of aspirin overdoses.

During the pandemic, aspirin was widely used, but many physicians seemed oblivious to its dangers. In Delhi, senior physicians were concerned that younger doctors in Bombay and Madras were misusing the drug, while in London one physician who practiced from Harley Street, London's fanciest medical address, encouraged its use. He recommended that the patient be "drenched with aspirin at a dose of twenty grains an hour for twelve hours, and then every two hours thereafter." That is six times greater than the maximum safe dose. It's an insane amount of aspirin.

Because it was given in highly toxic doses, it may have been the aspirin that killed so many during the pandemic, and not the influenza itself. This is an unsettling thought, but it might help explain the deaths of a disproportionate number of otherwise healthy young adults—the very population that today rarely suffers from serious flu infections.

Hopkirk also suggested that for pneumonia, a patient take "a teaspoonful of Friar's balsam, or a small handful of eucalyptus leaves" with a pint of water. Friar's balsam, in case you were wondering, contains benzoin, a resin that comes from the bark of several different trees. I used benzoin all the time in the emergency department; I'd dab it around a wound before placing a dressing over the top. Benzoin makes the dressing stick much better. But it's of no use in treating the flu.

Hopkirk, like many physicians of his day, also prescribed quinine to treat the flu.

"In quinine," he wrote with great certainty, "we have a drug that not only controls fever-producing processes allied to fermentations, but also exerts a definite anti-toxic action on the specific virus of influenza itself."

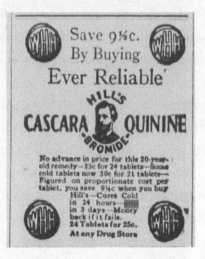

Vermont Digital Newspaper Project

Bark again. Quinine comes from the cinchona tree, found in South America. There it was used by the indigenous peoples to treat malaria, and by the middle of the seventeenth century it was imported into Europe, where it became known as Jesuits' powder (named for the religious order that brought it to Italy). Until only a decade ago, quinine was used as the first-line treatment for malaria, and it continues to play an important role in eradicating that disease. So how did it end up being used to treat influenza?

The answer is simple. Malaria causes fevers, just like influenza does, and quinine reduces the frequency and severity of fevers. If quinine cured the fever associated with malaria, why not use it to treat all fevers? And so quinine became a standard weapon in the arsenal to fight influenza. At the outbreak of the pandemic it was used in England, the United States, and the European continent. Grove's Tasteless Chill Tonic was the most popular quinine product. As a treatment for malaria it had made Edwin Wiley Grove rich in the 1870s, and now it was being marketed for the flu. In newspaper ads across the country, it was claimed that the tonic would "fortify the system against colds, grip and influenza." It improved the appetite, brought color to the cheeks, restored vitality, and purified the blood, "making it rich." You would soon feel its "strengthening invigorating effect" and, as an added bonus, Grove's tonic did not upset your stomach, or cause "nervousness or ringing in the head."

But quinine does not directly reduce fevers in the way that aspirin does, and so it had no effect on the fevers caused by influenza. Worse still, in high doses it causes vision problems or even blindness, ringing in the ears, and cardiac arrhythmias. Overall, quinine is a dangerous and useless drug for influenza.

Still, it wasn't all toxic mercury and tree sap for Hopkirk's hapless patients. For nausea and vomiting, which are common symptoms of the flu, he recommended small doses of dry champagne.

"There is no finer pick-me-up after an attack of influenza," he wrote, "than good 'fiz.'"

If this all sounds a little much, it was. Even a hundred years ago, the medical community thought Hopkirk's advice was peculiar, at best. An anonymous reviewer writing in the *Journal of the American Medical Association* could not hide his contempt:

> Foreign physicians, especially British, may find such a book tolerable and perhaps instructive, but for Americans the ordinary text-books will probably furnish an equal quantity of useful information without induction of nausea by the persistent recommendation of nostrums. The astonishing thing is that Scribner's would allow their imprint on such a book.

Astonishing indeed. But Hopkirk's remedies were not as unusual as you might think. In fact, they appear to have been pretty mainstream (even in America, much to the chagrin of that cranky reviewer).

One of my favorite examples of how we fought the flu comes from the 1936 nursing records of an influenza patient, which were saved as a family heirloom and published seventy years later. Over a period of three weeks he was treated with a punishing battery of balms: mustard plaster (a home remedy rubbed on the skin), aspirin (for fevers), codeine (for cough), phenolphthalein (a cancer-causing laxative), cough medicine, camphorated oil, seven enemas (seven!), rectal tubes (don't ask), milk of magnesia (another laxative, God help him), urotropine (a bladder antiseptic), and tincture of benzoin. The patient received at least five prescribed doses of whiskey and fourteen doses of castor oil. Actually, his seven enemas may have been medically necessary, because he was given at least thirty-nine doses of codeine, which suppresses coughing, but also causes constipation.

Now remember, this was two decades after the great influenza pandemic, and yet patients were *still* being treated with Friar's balsam and castor oil. What we can conclude from Hopkirk's 1914 book—and the nursing records of this poor overtreated patient—is that doctors attacked influenza with a number of folk remedies that were at best useless and at worst poisonous.

Some were at least organic: burning orange peels, or dicing onions to sterilize a room. Many doctors had concocted potions and medicines of their own, and advocated for their use with statistics that are hard to believe. In February 1919 a Dr. Bernard Maloy of Chicago claimed that he had treated 225 patients with pneumonia, *and had not lost a single one*. He used a tincture of two plants, aconite and *Veratrum viride*, in a regimen of ten doses. We do not know the concentrations of each ingredient, but aconite (also called monkshood) and *Veratrum viride* (called false hellebore or Indian poke) are plants that are—you guessed it—poisonous. In sufficient doses they cause nausea, vomiting, and a precipitous drop in blood pressure. They may even be fatal. Maloy's mixture must have been carefully titrated to prevent these side effects, and let us not forget that many modern drugs are also toxic in high doses. Still, his claim that the mixture prevented or perhaps aborted pneumonia with a success rate of 100 percent suggests that his patients had been carefully selected, and that those with severe cases of influenza or pneumonia had been excluded from his protocol.

Some people were so desperate during the 1918 pandemic that they found their own perilous means of treatment, without the aid of a misguided doctor. As the flu roared through the coastal towns of southwestern England, the villagers of Falmouth were taking their sick children not to the hospital but to the local gasworks—to inhale the fumes. Parents thought that exposing their children to poisonous gases would reduce their symptoms.

A sanitary officer, Captain A. Gregor, set off to investigate this claim scientifically by looking at the rates of influenza across different groups in Falmouth. At a naval patrol base, he noted a 40 percent rate of influenza. A local army battalion of 1,000 troops had a rate less than half of that. And at a local tin works where workers were exposed to the noxious fumes of nitric acid, the influenza rate was half the rate at the army battalion: a mere 11 percent. Some workers at the tin works were exposed to explosives and gunpowder, and for those lucky enough to inhale these fumes, the influenza rate was lower still; only 5 percent of them came down with influenza.

The popular belief that many "colds in the head" could be cured by fumes "has some foundation of truth," Gregor concluded in the *British Medical Journal* in 1919, as the influenza pandemic petered out. He was not the only one to make this observation. Another physician reported that it was "abundantly clear that poison gas workers were practically immune from influenza." Mercifully, no one—not even the mercury-friendly Dr. Hopkirk—was actually recommending breathing toxic fumes as a way to prevent influenza.

There is no way of knowing if Gregor's observations really had anything to do with the exposure of the workers. Chlorine does kill the avian flu virus, and it probably did the same to any pandemic flu viruses floating around those gasworks, but remember, chlorine gas was also used to kill scores of soldiers in the most excruciating way during World War I.

Not all physicians practiced like quacks during the pandemic. James Herrick, a doctor who worked in Chicago, studied medicine at Rush Medical College in Illinois, and was by all accounts a brilliant physician. In 1910 he was the first to describe what was later called sickle cell disease, although at the time he was stumped to explain the cause of the condition. Two years later he published an important review of diseases of the coronary arteries in which he argued, against the prevailing wisdom, that those arteries may become blocked without causing immediate death. Based on his experiences, he was able to describe the clinical manifestation of such blockages a century before cardiac imaging became available. In doing so he laid the foundations of modern cardiology. In addition, he published on pneumonia, leukemia, and a host of other diseases, including influenza.

Herrick was among the very first to challenge the potions and folk remedies that were actually harming and killing patients who had influenza. Herrick had treated the flu during two pandemics, 1890 and 1918, and his plea was simple: doctors should stop using nearly all the medications in their arsenal. There was no evidence that any of them worked.

It took a lot of courage to write that in the summer of 1919, when the United States and the rest of the world were recovering from the worst pandemic in history. Most doctors who treated the flu, Herrick wrote, did so on the basis of "superficial observation and limited experience." They ignored the fact that the disease is self-limiting, meaning that it usually cures itself.

"So many conclusions are crude," Herrick wrote, "and so many are reached by a mental process in which an optimistic credulity takes the place of the more desirable scientific skepticism."

Herrick balked at the variety of bogus treatments, which ranged from the inane to the lethal. A shot of mercury? Enormous doses of quinine? "Certainly," he wrote with a characteristic understatement, "some one has blundered in reaching conclusions."

Instead of prescribing these worthless medications, Herrick said, let's try more practical methods that really do work: isolation, for example, and masks to prevent infection, with plenty of fluids to keep the patient hydrated. And rest. Lots and lots of rest. His regimen was squarely in the conservative mainstream. Several weeks of bed rest, partly outdoors, and plenty of fresh air, quiet, and sleep.

Of course, Herrick was also a man of his time, so it's not surprising that he, too, addressed the use of laxatives, and insisted that "the bowels should be opened fully at the beginning of the illness and not allowed to become sluggish at any time." But let's give him a pass on that one, because of all the other marvelous and commonsense things he had to say:

One of the hardest things to do in the treatment of a serious, self-limited, infectious disease is to refrain from prescribing drugs merely because the diagnosis has been made. The self-restraint of the level-headed physician is likely to be swept aside by the thought of the possible grave consequences of the malady, and his accustomed good judgment is apt to be smothered in the semihysterical atmosphere of alarm that pervades the community during the visitations of the epidemic. He forgets that a large proportion of patients with influenza do not need a single dose of medicine. There should be no routine treatment according to which certain drugs are given at stated periods, whether or not there is a clear indication for their use. The treatment is really expectant, symptomatic and individualistic.

That last sentence is gold. It should be ingrained in the brain of every medical student in every medical school in the country. Wait and see what happens, treat the symptoms, and think about your patient and the individual profile she fits.

Fortunately, there were other physicians who also believed that the vast majority of the interventions for flu were flawed, at best. In November 1918, one doctor stationed with the Canadian troops at Camp Bramshott in England wrote that from the large number of agents used to treat influenza, "their comparative futility is obvious."

While treatments have changed over millennia, and indeed from decade to decade, the patient in some ways remains the same. It's the same type of virus, after all, that afflicted the ancient Greeks, that sent unlucky souls to Dr. Hopkirk, and that still knocks out your spouse, your child, or yourself. So what now?

Well, none of my colleagues will offer you a laxative, at least. We won't order bloodletting either. But you might be surprised at how little the treatment of influenza has advanced.

ENEMAS, BLOODLETTING, AND WHISKEY: TREATING THE FLU

Here is a typical rundown of what happens more than 31 million times each year in the United States. It is late fall and you start to feel unwell one Friday evening. You are tired and don't feel like eating. Your lower back and legs start to ache. Then you get an episode of chills and begin to sweat. You take your temperature. It's 102 degrees Fahrenheit. Now you really start to feel awful. The chills get worse. Your throat gets scratchy, then sore. You start sneezing. By Saturday morning you have a runny nose and a cough, and now your entire body aches. You have influenza.

How you react to this very common scenario varies. You may stay home and reach for Tylenol or Motrin to keep the fever down and ease the aches. You may stay in bed, drifting in and out of sleep. If you are lucky, you've got someone to check in on you and bring you water or a hot drink. After a couple of days, the fever finally abates, and your strength starts to come back. It's now Monday, so you call in sick, but you can finally drag yourself into the shower. Although you have little appetite, you drink some soup. By Tuesday your fever has gone, and your appetite slowly returns. It's all over by Wednesday, and you are back in the office.

That's what happens to most otherwise healthy people when they get the flu. Most. But certainly not all. At the first sign of a fever or body aches, some will call their primary care doctors, who will tell them to stay home and drink plenty of fluids, or to go to the emergency room if their condition does not improve. The last thing your doctor wants you to do is come into her office and infect her, her staff, and the other patients. I've treated hundreds of patients with the flu in the ER, and many were there in the very early stages, or with symptoms that were so mild that all I could do for them was send them home with the kind of advice my mother would have given: have some chicken soup.

Some patients, though, are in mortal danger when they get the flu. They may be elderly or have an immune system compromised by HIV, chemotherapy, or steroids. They may have an immune system that functions perfectly well but they just happen to have a particularly nasty bout of influenza. They may not have drunk enough water, or they couldn't stay hydrated because of vomiting and diarrhea. These are the more serious cases of flu, the ones that often end up in the ER. Most arrive by car or cab, and some by ambulance.

However you get there, the first person you'll meet on arriving in the ER is a triage nurse. She'll ask you for a quick medical history. Then she will take your pulse and blood pressure, check your temperature, and put a little probe on your finger to measure the oxygen content of your blood. If these four measures, known collectively as your "vital signs," are more or less normal, you'll be given a mask to cover your mouth and nose and asked to sit in the waiting room until a bed opens up. As you sit there, you may see three or four other masked patients, also in their pajamas with overcoats draped over their shoulders, waiting, just like you. The sickest patients go into the ER first; if you can stand but another patient is too weak to walk, he gets moved in front of you.

If the flu season is especially severe, there will be many patients with symptoms just like yours, clogging up the waiting room. If you arrive in the afternoon or early evening, peak times for most ERs, your wait will be longer. If you are treated in an urban ER, you will likely spend longer waiting than if your ER were in the suburbs. Fridays and Mondays are often the busiest days of the week; federal holidays and the early-morning hours are usually quiet. On the day after a federal holiday, the ER is terribly busy. Remember, too, that medical teams are likely to be at their slowest at the end of their shifts. Putting this all together means that if you have a bad case of the flu and need to be seen in the ER, your best bet is to turn up at seven a.m. on Christmas morning. Just don't tell them I sent you.

Once a bed opens up for you, you'll be poked and prodded. An IV goes into your veins. Blood samples come out. All before a doctor sets eyes on you. When the doctor arrives, she will ask you about your illness: start time, symptoms, and so on. The doctor has two goals. First, she wants to make sure that you don't have a serious condition like pneumonia that could require antibiotics or admission. Second, she wants to figure out if you require any interventions, like additional intravenous fluids. If you do indeed have the flu, and you don't need IV fluids, you will be sent home with nothing more than some Tylenol (and, in the U.S., a rather large bill).

So how does the doctor know that you actually have the flu? I have to admit that even after five years of medical school, another four years of residency, and several thousand hours of seeing patients, most of us in the ER just intuit it. Of course, we rule out other conditions by asking important questions like "Have you traveled to Africa?" or determining whether you've had any exposure to carbon monoxide. That last question is really important. If it doesn't kill you immediately, carbon monoxide poisoning will cause symptoms that mimic the flu. Since flu outbreaks peak in the fall and winter—the very period in which people are running their heaters and furnaces—carbon monoxide exposure is often misdiagnosed as the flu.

Several years ago, I appeared in court as an expert witness in a tragic malpractice claim in which a husband, wife, and son were found dead in their Philadelphia home from carbon monoxide poisoning. It turned out that the wife had visited her local ER with headaches, nausea, and vomiting. Twice. On neither occasion was the question of carbon monoxide poisoning considered. Instead, her symptoms were assumed to be caused by the flu. The jury awarded the estate nearly $1.9 million in damages.

Once a flu diagnosis has been made, the discussion of treatments begins. If you have a fever, you will be given a medicine to lower it. That's what every ER doctor in the country will do, including me. But it's good to ask if we *should*, in fact, reduce the fever associated with influenza.

For nearly everyone, fevers are not dangerous in any way. But they are rather unpleasant, and so we treat them. There is evidence that a fever is actually beneficial, and the reason is simple: the immune system fights infections better when the body is hotter. White blood cells are released in greater numbers from the bone marrow, and the cells do a better job of fighting the infection. Fever also improves the efficacy of another group of blood cells called natural killer cells, and it increases the ability of macrophages ("big eaters" in Greek) to ingest and destroy the invading cells.

Since the body does a better job of fighting infection when it is a few degrees hotter, might reducing the fever lead to a *worse* outcome for the patient? A group from McMaster University in Canada looked at what happens in a large group of people when some of them—infected with, say, influenza—take medicine to reduce their fever. Once they feel better, patients with the flu get out of bed and start to socialize, spreading the virus. On a population level the effect is rather drastic. The McMaster group concluded that the practice of frequently treating fevers with medication enhances the transmission of influenza by at least 1 percent. I know that doesn't sound like a lot, but remember that as many as 49,000 people die from the flu each year in the United States. If you plug the McMaster estimates into these flu numbers, almost 500 deaths per year in the U.S. (and perhaps many more elsewhere) could be prevented by avoiding fever medication during the treatment of influenza.

In the ER, I would always give medicine to a flu patient with a fever. And so, I believe, would every ER physician I know. It's partly because that's how I was trained, and partly because fevers are just not pleasant. But it's also because of patient expectations. People expect their fever to be treated. It's just not worth the time and effort to explain the McMaster paper to a sick and achy patient who desires relief.

Another intervention I often provide for flu patients is intravenous fluids. For dehydrated patients this is extremely important. After a bag or two of IV fluid—which contains sterile water, salt, and a bunch of electrolytes—patients often feel remarkably better. I've seen countless patients with the flu arrive at the ER by ambulance, too weak to stand. An hour later, and with two bags of fluid on board, they are able to walk out of the ER and return home.

Blood tests are usually not necessary, and a chest X-ray exposes the patient to needless radiation. This is important for two reasons. First, because you may be the kind of person who arrives at the ER with a mild case of the flu and expects the doctor to order blood tests and an X-ray. And second, because if you're not that kind of person, you may find it hard to believe that there are indeed those who want this kind of testing done as a matter of routine. Leave it up to your doctor. Don't request a blood test or X-ray. They serve no purpose other than to add a large charge to your bill. I almost never order them, but there are exceptions. Some patients will just look very sick, be extremely dehydrated, or have complicating chronic conditions. Some may be heavy smokers, and some may have developed pneumonia. They might be out of breath. When I listen through my red stethoscope to their lungs, I hear crackles and wheezes (or "rales" and "rhonchi," as doctors mysteriously refer to them). In these patients, a lung X-ray is vital, because it will show if there are signs of pneumonia. A blood test will show high numbers of white cells, signaling a serious infection. One of the first ways I can help is to give such patients pure oxygen to breathe, through a clear plastic mask placed over their nose and mouth. In our lungs are thousands of tiny sacs called alveoli, through which oxygen passes into our bloodstream. In lungs that are ravaged by influenza and pneumonia, these alveoli are filled with fluid and pus. That means less oxygen is getting into the blood, leading to shortness of breath. Blood with enough oxygen is bright red. Blood without it turns darker. When the oxygen levels get critically low, the lips and ears turn a dusky blue. This is called cyanosis, and it's a sign that the patient is very sick. It was one of the hallmarks of severe illness in the 1918 pandemic. To treat cyanosis or a low oxygen level, I give pure oxygen. This can relieve distress in just a few minutes.

These sick patients must then be admitted to the hospital. They will be treated with antibiotics to fight the bacteria that have taken hold in the lungs. They will get IV fluids to keep them hydrated, and will continue to breathe pure oxygen that is fed to them through plastic tubes attached to the wall. Most need only a few days in the ward to improve, but if the damage to the lungs is severe and widespread, they will be transferred to the intensive care unit. There, each patient is looked after by a single nurse, attentive to every change in their condition. If their illness gets even worse, they will be sedated and connected to a machine that takes over the work of breathing. A tube about nine inches long and the width of your index finger is slipped carefully into the trachea and past the larynx. It is attached to a ventilator, and with each cycle the patient's chest expands and contracts. Now all we can do is wait.

If all goes well, the pneumonia is beaten into submission and the inflammation caused by the influenza virus slowly recedes. After a few days, the breathing tube can be removed and the sedation is slowly reduced. The patient awakens, oblivious to the life-and-death battle that has been raging. That's if things work out. But sometimes the pneumonia is so virulent and aggressive that it cannot be stopped. First the lungs will fail, then the kidneys and the liver. Multiple organ failure. And influenza takes another life.

I don't mean to be too morbid here. Of the millions who typically come down with the flu each year, fewer than 1 percent will die. For those who visit the ER, most need only to be reassured that time is all that is necessary to heal. One of the great current misconceptions is that antibiotics are needed for all ailments. If you are a healthy person with run-of-the-mill flu, you should not ask for antibiotics, and your doctor should certainly not prescribe them. Antibiotics don't fight viruses, and so they are completely ineffective against the flu. If, however, you have a complication and your viral flu has evolved into a bacterial pneumonia, you should certainly be treated with them. But, and it's worth repeating, antibiotics do nothing against the flu virus. You would be amazed at how many patients ask for antibiotics when they clearly have a viral infection, and are disappointed when I decline to prescribe them. Doctors are largely responsible for this huge problem. The best data we have suggests that about half of all patients with viral infections like influenza get a completely useless antibiotic.

Bloodletting, enemas, champagne, toxic fumes, and castor oil. It's hard to believe we once thought these were state-of-the art treatments for the flu. We've come a long way over the last hundred years. Or so you might think. But despite all the benefits that modern medicine provides, curing the flu remains beyond our reach. We are still threatened by the virus and worry that another pandemic like 1918 is just around the corner. To understand why a cure remains so elusive, we need to take a closer look at the virus itself.

2

THE JOLLY RANT:
A HISTORY OF THE VIRUS

Viruses were here long before us. They were here before intelligent life, before apes and chimpanzees, before reptiles, before anything crawled out of the slime of creation. Viruses are ubiquitous, but their mystery is inherent in their origin: we don't know exactly how they developed, but we do know they have been around for millions of years. They exist on the edge of life, and they challenge our definitions of what it means to be a living thing. A rock is not alive, but a bacterium is. Viruses lie somewhere between the two.

A virus is a box of chemicals, without the structures of a basic cell. It cannot metabolize or replicate on its own. In order to reproduce, it must invade living cells. Viruses infect bacteria and plants, reptiles, fish, birds, and mammals. They have been so intertwined with our own evolution that over millennia some became incorporated into our own genetic code. Nestled into the long strands of our DNA are sequences that origi- nated from ancient viruses. Their genetic code was so bound with ours that the virus became a harmless part of us, freeloading off human cells as they reproduced.

*   *   *

The word "virus" was used long before the discovery of the particles that now bear this name. It's a Latin word meaning "poison," "venom," or "noxious smell." In the Middle Ages "virus" was a synonym for "toxin," and the word remained untranslated in the English versions of Latin medical texts. By the eighteenth century it meant any infectious disease; Edward Jenner, for example, used the word to describe the cause of smallpox before he discovered a vaccine to prevent it. In the nineteenth century, as the germ theory of disease was rapidly developing, "virus" was still used to denote any infectious agent, bacterial or not. Louis Pasteur referred to the agent that caused rabies as "*le virus rabique*." Today, we know that viruses are submicroscopic entities twenty times smaller than a bacterium. They contain a core of genetic material covered by a protein capsule, and they reproduce exclusively within living cells.

Just as the word "virus" was used long before it gained its present meaning, the word "influenza" arose long before its current use. No one is sure if "influenza" was first used in the English language to describe the same disease that now bears this name, but it appeared as early as 1504. It comes from the Italian word meaning "influence," which attests to the astrological theory of its origin. We once thought that influenza was caused by a misalignment of the stars and planets.

We didn't know precisely what a virus was until the twentieth century, so humans had millennia to fret over and hypothesize about this unseen force. The ancient Greek historian Thucydides, who compiled a history of the Peloponnesian War between Athens and Sparta, wrote of a three-year plague that struck in 430 BCE. Tens of thousands of refugees had entered Athens seeking shelter and protection. The city quickly became overcrowded. These were the optimal conditions for the outbreak of an infectious disease. The first symptoms were what Thucydides described as "an overwhelming fever in the head and redness in the eyes," followed by sneezing and hoarseness, which "in short order settled in the chest with a violent cough." The fever was so severe that the afflicted threw themselves into water cisterns, where they would cool off and drink to assuage their unceasing thirst. Thucydides was surprised at how long the victims clung to life, but most died within a week. The epidemic killed one-third of the 13,000 soldiers stationed in Athens. Then, strangely, in the winter of 427 BCE, it came to an abrupt and welcome end.

This disease has long been a point of historical intrigue. Plague and typhus were the usual suspects, but other theoretical culprits included anthrax, typhoid, and tuberculosis. It appeared as a sudden outbreak and had a short incubation period. Those who were sick and recovered—including Thucydides himself—did not become ill a second time. The disease came in waves and was spawned by overcrowding. In the 1980s, researchers called this group of symptoms "Thucydides syndrome." They also noted that the symptoms were characteristic of pandemic influenza complicated by a secondary bacterial infection. The outbreak shared many features with the 1918 flu epidemic, including secondary infections that led to most of the deaths. If this theory is correct, Thucydides syndrome is the earliest record of epidemic influenza. Given the incredibly high death rate, it was also the most lethal.

A century after Thucydides, the Greek physician Hippocrates wrote about an annual outbreak of a disease that also sounded like influenza. It aligned with the appearance of the star cluster known as the Pleiades, or the Seven Sisters, which in the Northern Hemisphere is seen in the fall and winter months. During this time, wrote Hippocrates, there were "continual fevers that attacked the population in great numbers." The sick had chills and frequent sweats accompanied by coughing.

After that, there were no records of influenza outbreaks until the late Middle Ages, by which time smallpox and bubonic plague were the most feared and deadly diseases. Compared to these mass killers, influenza was barely noticeable.

Centuries later, in November 1675, a flu epidemic broke out in my home city of London. The number of deaths per week grew from 42 at the beginning of the month to 130 by its middle, before falling to only 7 deaths in the first week of December. Aside from actually killing people, there were other inconvenient features. Church congregants were coughing so much that it became impossible to hear the sermon. In perhaps a fit of irony, Northern Englanders called the disease the "jolly rant" because it turned its victims into miserable noisemakers. There was nothing jolly about it, of course. The famous seventeenth-century English physician Thomas Sydenham suggested these epidemics were related to heavy rains that filled the blood with "crude and watery particles." Bloodletting and laxatives were thought to be the best cures.

Let's digress from blood and bowel movements to distinguish *epidemics* from *pandemics*. These words were—and are—frequently used interchangeably to describe outbreaks of influenza. In 2009 there was an influenza outbreak that became known as swine flu, and it was a perfect example of the confusion of the two terms. "Is This a Pandemic? Define 'Pandemic'" read a headline in the *New York Times*. No one really agrees on exact meanings, though scope and intensity are at play. The most useful definition we have is that an *epidemic* is a severe local outbreak, while a *pandemic* is a global outbreak that makes people very sick, and spreads rapidly from a point of origin. Using this rubric, there were three to five flu pandemics in each of the seventeenth, eighteenth, and nineteenth centuries. Some of these outbreaks were separated by as much as half a century, while others appeared within a few years of each other. This is partly what makes the flu so confounding: it is predictable on a small scale, season to season, but unpredictable on a large scale, in terms of epidemics and pandemics. For example, the influenza outbreak of 1730 was followed two years later by a sequel. Almost exactly one century later, another one-two punch occurred in 1831 and 1833. At other times, a respite might last for fifty years. Influenza took so long to track and identify because it was so unpredictable.

One particular nineteenth-century pandemic stood out from those before it and brought us closer to solving the mystery of what was out to get us. The devastating outbreak in the winter of 1889 was not only the first influenza pandemic in modern times but also the first that was sufficiently detailed and documented, allowing its spread and impact to be assessed. It was the first influenza pandemic in Britain in more than forty years, and it was so severe that a doctor named Henry Parsons reported on it to Parliament. Parsons noted that the outbreak was certainly a pandemic: all of Europe had been afflicted. The disease spread to the United States, where the first cases were reported in New York in December 1889. By January there were deaths in Boston, St. Louis, and New Orleans. In Boston, 40 percent of the population became ill. More than a quarter of all laborers were too sick to work. Overcrowding and deadly "impure air" had a powerful effect. Rich and poor suffered in the pandemic, but as expected, it had a greater prevalence "among persons associated together in a confined space."

Parsons was at a loss. He couldn't suggest ways to prevent influenza, because there was still a central mystery: its cause. It was anyone's guess. The Parsons report to Parliament suggested that the pandemic had begun in Russia and then spread westward. But how much of this was scientific analysis and how much was jingoism? There was even a rumor that the pandemic had been brought to Britain by imported Russian oats, which were eaten by horses who then spread the infection into the human population. Other theories of origin included rotting animal carcasses, earthquakes, volcanic eruptions, and "effluvia" discharged into the air from "the bowels of the earth." It was even suggested that the pandemic was caused by the conjunction of Jupiter and Saturn.

Parsons suggested three possible origins for the influenza pandemic of 1889. The first was the weather. This would explain why so many cases of influenza arose almost simultaneously across Europe and America. Perhaps the air quality was poor. Or was it possible the atmosphere was carrying a poison that was able to multiply in midair and then infect a few susceptible people? Parsons admitted that he knew of no agent that could do this, though he suggested that it could be "non-living particulate material"—a remarkably accurate description of what a virus turned out to be.

The second origin theory was that influenza was transmitted from one person to another. This would explain why large numbers of household contacts were often infected together, and why in many instances a single member could be identified as having introduced the disease into a family. In a clever piece of detective work, Parsons obtained influenza data from the workforce of Britain's large railway system. Infection rates were higher among the clerks—who were not exposed to outside air but spent all day in contact with many people—than the engine drivers, who were essentially quarantined from the masses but out in the open more. Parsons was convinced that human interaction was a major culprit.

Parsons's third theory was that animals were somehow responsible—specifically horses, pet dogs, cats, and caged birds. Once again, Parsons was onto something, in this case about fifty years before everyone else.

Scientists were aware of bacteria before they'd nailed down what a virus was. By the 1840s, several European scientists independently arrived at the conclusion that yeast—the essential ingredient in the fermentation process—was a living organism. This meant that fermentation was not only a chemical process but also a *biological* one, caused by the activity of microbial organisms. The Frenchman Louis Pasteur studied the way fermentation relied on yeast and other organisms too small to be seen. His surname became identified with the process of heating liquids to kill bacteria (pasteurization). Pasteur, who was born in 1822, dabbled in chemistry before turning his attention to the problems facing local breweries in Lille, on the northern border of France. He showed that fermentation required not only the living yeast but also an additional kind of microbe, which he could observe under the microscope: bacteria.

Pasteur's discovery of bacteria changed the face of biology in general, and medicine in particular. Since at least the time of Aristotle, philosophers and scientists had believed that spontaneous generation explained the appearance of any number of biological phenomena. It was the reason why maggots appeared on rancid meat, why some plants could germinate without seeds, and why fungus grew on rotting fruit. But in a series of clever experiments in the 1850s, Pasteur showed that if an object was properly sterilized, spontaneous generation would not occur. By 1877 scientists determined that bacteria caused infectious disease. These microbes were soon identified by name. Anthrax was shown to be caused by a bacillus, a particular type of bacteria. Soon afterward scientists identified the specific germs behind throat infections, pneumonia, leprosy, and others. This explosion of identifying bacteria had an unintended consequence. In their zeal and eagerness, scientists identified microbes as the cause of a number of diseases when, in fact, they were nothing of the sort. Many of these bacteria were actually secondary pathogens that invaded a weakened host. They were associated with a disease but were not its source. This is precisely the mistake that was first made in identifying the cause of influenza.

In 1892 two microbiologists working in Berlin claimed to have discovered the bacterium that caused pandemic influenza. They called this new agent *Bacillus influenza*. Others referred to it as Pfeiffer's bacillus, after one of the microbiologists, Richard Pfeiffer. They were wrong, of course. The bacteria were certainly present in flu victims, just not as the cause of the flu. Instead, they were a secondary pathogen that invaded a body whose immune system was overwhelmed by what we now know to be *viral* influenza. The bacteria didn't *cause* influenza any more than circling vultures cause the death of a deer felled by wolves. Alfred Crosby, a historian of the 1918 outbreak in the U.S., described Pfeiffer's bacillus as "an authoritative road sign pointing in the wrong direction."

Today, the bacterium *Bacillus influenza* goes by a different name: *Haemophilus influenzae.* I've prescribed antibiotics to fight this nasty bacterium many times, but had not understood why its name included the word "influenza." It is the cause of pneumonia, meningitis, ear infections, and more—but never influenza, obviously. Once I'd schooled myself in the history of flu-related confusion, its misnomer made sense. It comes from a century ago, when our knowledge of influenza turned out to be wrong.

We've found the virus now, but what exactly *is* it? What is this thing that causes the phlegmy sniffles of the common cold, but can sometimes kill with the bloody violence of Ebola? How do viruses spread and afflict in so many different ways?

Viruses have evolved to be very different from the cells that are found in our body. Cells contain tiny specialized organs. Viruses have nothing similar. They have no mitochondria, so they cannot make energy. They have no ribosomes, so they cannot build proteins. They also lack lysosomes, which export waste and toxins. The virus is simply an envelope containing a bundle of genes that exist only to make copies of themselves. But while a computer virus is designed to infect your laptop and cripple or compromise its functions, nature's viruses do not target cells expressly in order to kill them. Instead, their singular purpose is to hijack a cell and use it as a copying machine. In doing so, they may injure or destroy the host cell, but that is collateral damage, not their primary objective. In fact, if a virus is too lethal, it can kill the cells before they can be used to make viral copies. The influenza virus, the human immunodeficiency virus (HIV), and the Ebola virus differ in how deadly they are, but their strategy is the same. They invade our cells, reproduce, and then must find a new victim to colonize. They may leave their host weakened or even dead, but this is incidental.

We now recognize more than two thousand species of viruses, and the number keeps growing. Most doctors are familiar with just a few of them. There are the herpes viruses, which give us chicken pox (and, well, herpes). There are the rotaviruses, which cause diarrhea in young children. There are about a hundred different rhinoviruses, responsible for the common cold, and there are retroviruses, like HIV, which causes AIDS. We are especially interested in one particular family of viruses that has the clumsiest name: the orthomyxoviruses. *Ortho* means "straight," in Greek, and *myxa* means "mucus." The straight-mucus family of viruses includes influenza. Actually, there are three influenza viruses—they go by A, B, and C. Only the A and B strains cause significant disease in humans, and it is the A strain that is responsible for pandemic flu.

Influenza is an incredibly simple virus to depict. It is shaped like a hollow ball and contains eight viral genes, made up of RNA (used in place of DNA), which control the function of the virus.

Sticking out of its exterior are two important proteins that look like tiny spikes and pitchforks. The spiky protein is called hemagglutinin, or HA for short. After the virus is inhaled into our lungs, it is the HA that latches on to the surface of our cells. Now the virus has one foot in the door. The cell is tricked into absorbing the virus, and once inside the cell the viral membrane dissolves and releases its eight genes. These enter the nucleus of the invaded cell. From there they commandeer the normal machinery and direct the cell to make millions of copies of the viral particle. These baby particles then rise back to the cell membrane, like bubbles in a boiling pot. At the surface, they become tethered and must break free in order to invade other cells. This is where the second, pitchfork-shaped protein on the surface of the flu virus steps in. It is called neuraminidase, or NA. Its function is to break the bond between the surface of the cell and the surface of the virus. The viral progeny is now free to be coughed or sneezed out, to invade another victim. The whole process takes only a few hours, and it leaves behind destroyed respiratory cells. That is when the symptoms of influenza begin.

During the process of reproduction, the flu virus may change in one of two ways, and as a result of these changes new strains of the virus appear. The first occurs when there are copying errors in the instructions to build a new virus. These instructions are stored on the eight viral genes and are built from the genetic code. When the virus reproduces, that code is read and copied millions and millions of times. But the copying process is not perfect. Reading or manufacturing errors occur. As a result, the code in the progeny virus can differ from that of the parent virus from which it was copied. These differences in the genetic instructions result in slight modifications in the surface proteins on the virus. Since our immune system learns to identify a flu virus by the proteins on its surface, the modifications result in an unrecognizable flu virus. That's how new strains develop, and why we may become ill with the flu many times. In essence, we are being infected with a new virus each time.

To understand the second way in which new strains can occur, we must understand that influenza A is not just found in humans. It infects many different species, like pigs, birds, and horses. Sometimes, two or more different strains of the virus invade the same lung cell. There, the genes from each strain can mix together and produce a hybrid virus, containing the genetic material from both parents. Mammals get the flu in their lungs, while in birds the virus is found in the gut. An infected bird's droppings can contain billions of avian flu viruses, each ready to mingle with the genetic material of other strains of influenza, including those that infect us. If a bird strain and a mammalian strain of influenza invade the same cell at the same time, their genes can mix to create an entirely new kind of flu virus that can turn deadly. That's what happened in 1918, when birds contributed to parts of the influenza virus that nearly brought humanity to its knees. It also happened in Hong Kong in 1997. A new avian influenza virus infected those who worked closely with chickens. It killed six of the eighteen people in whom it was identified. The virus could infect only those who directly handled birds and could not be transmitted from one person to another. But it would take merely a small mutation for the virus to gain that ability, setting the stage for a new influenza pandemic.

Although a single flu virus can produce millions of descendants as it takes over a cell, only a small number are actually able to reproduce. Nearly every genetic change that occurs results in a viral particle so damaged that it loses its ability to replicate. But given the millions of viral particles that are generated when the flu takes over, even a success rate of 1 or 2 percent can result in thousands of new flu viruses that can burst out of the cell and infect others.

Our immune systems have evolved to prevent and contain infections that viruses, bacteria, and other foreign pathogens may bring. The first line of defense consists of cells called phagocytes, whose name comes from the Greek meaning "devouring cell." Phagocytes are kind of like traffic police. They're always on patrol. They detect pathogens, envelop them, and pull them inside the cell. Once there, the pathogen is obliterated. Phagocytes do not target specific bacteria or viruses. Rather, they have been programmed in our genetic code to recognize pathogens in general. We are born with this innate immunity, and the phagocytes require no prior contact with a pathogen to search for, recognize, and destroy it.

The second line of defense in our immune system are the antigen-presenting cells, which target specific viruses or bacteria. These cells are like detectives; they profile a suspect. They digest the pathogen and present some of its building blocks—a protein, for example, or a receptor—to another kind of immune cell called a helper T cell. These T cells then proliferate in vast numbers and use the pathogen profile to target the corresponding enemy. Even years after a first encounter with a pathogen, T cells remember their old foe and spring into action. That's why most of us get chicken pox only once. Our first encounter with the virus produces T cells that are forever on guard in the future.

Our bodies learn to defend against new invaders all the time. Vaccination takes advantage of this by presenting our immune system with a weakened or harmless form of the pathogen. This allows us to manufacture antibodies in advance of an infection. The immune system doesn't care whether it came across the pathogen in the normal course of events or the pathogen entered the body as a vaccine via a needle. Either way, the immune response is the same. The body is able to fight an infection more quickly and effectively the next time around. If the antigen has not been previously recognized by our immune system, we may still produce antibodies against it, but the process is slower. We become sicker for longer, and our inability to mount an immediate attack on the virus may in some cases be fatal.

Influenza throws a wrench into our elegant system of defenses because it is a shape-shifter. It frequently changes the proteins on its surface, making it harder for our body to recognize them. Think of a criminal who makes convincing disguises, who can easily disappear in a crowd. These alterations provide the virus with an invisibility cloak, making it unrecognizable by existing antibodies. This is why you may catch the flu more than once during the season: your body produced antibodies to the first virus, but is infected by a second that it did not recognize. This "antigenic drift" is also the reason that the flu vaccine needs to be updated every year. The enemy is constantly dyeing its hair, so to speak, or wearing different masks.

In addition to antigenic drift, there is a larger change that the flu virus can undergo, antigenic *shift*, and that's how we get our flu pandemics.

During this shift, viral proteins assume an entirely new structure, and the virus is said to be "novel." These novel viruses—which most often arise when animal and human viruses share and swap their genes—are like entirely new criminals, not old ones in disguise. This makes them sneakier, more prolific, and perhaps more deadly. Antigenic shift generated the deadly 1918 influenza virus and the swine flu outbreak of 2009.

Through drift, shift, and sharing genes, influenza can morph faster than our bodies can profile it. During the time that the immune system takes to produce antibodies to one strain, a different influenza strain can take hold and turn deadly. The flu virus has perfectly evolved to stay one step ahead of our immune defenses.

The novel virus of 1918 killed tens of millions. The first reports came from Europe. A medical report in June of that year was short and mostly vague, but it was clear about the location of the outbreak:

A disease of undetermined nature was reported present, May 28, 1918, at Valencia, Spain. The disease was stated to be characterized by high fever, to be of short duration; and to resemble grippe. A number of cases similar to those which were reported at Valencia have been notified in other cities in Spain.

The next month, buried among news of the fighting in Europe, the *New York Times* noted that a new disease, the "Spanish influenza," was "now epidemic all along the German front . . . hampering the preparations for offensive operations." No one was immune. Within a month, the German Kaiser himself would catch it. Like a well-trained military, influenza seemed to have its own tactics and strategy. It was stealthy. It attacked on all fronts, and not just once. And its first victims were soldiers who had expected to be fighting an entirely different kind of battle.

3

"SOMETHING FIERCE":
THE SPANISH FLU OF 1918

Dr. Loring Miner was a country doctor in rural Kansas who practiced an entirely different kind of medicine from that which we have today. He lived far from the nearest hospital, at a time when the machinery of modern medicine was simply unimaginable. Despite the technological limitations of his era, Dr. Miner played a critical role in the story of the 1918 epidemic.

In 1918 Miner had a vast office. His rural medicine practice covered 850 square miles of flat farmland that was nurtured and harvested by 1,720 potential patients. Haskell County, a perfect square of land in southwest Kansas, was 200 miles west of Wichita. In January and February of 1918, as farmers hunkered down in their houses, Dr. Miner observed dozens of cases of severe influenza, or what he called "a disease of undetermined nature." In one day alone, eighteen people fell ill. Three died. In a sparsely populated county like Haskell, this was remarkable enough to prompt Dr. Miner to write a report to health officials. It was the first recorded instance of a physician warning about an outbreak of influenza. We are far from sure, but Haskell County might have been ground zero for the 1918 flu epidemic in the United States, and perhaps in the world.

Three hundred miles to the east was the U.S. Army's Camp Funston. Soldiers from the camp visited family in Haskell County at the height of its influenza epidemic and returned to base in late February 1918. On March 4, the first soldier at Camp Funston fell ill with influenza. As soldiers moved freely between Funston, other army camps, and the civilian world, the virus expanded outward in waves. It reached France first at Brest, the largest point of disembarkation for American troops, and on it spread. These facts support the hunch (but it's only a hunch) that the worldwide flu pandemic of 1918 originated in the American heartland.

The evidence points to two other possible ground zeros. The first was France. John Oxford, a virologist from the University of London, noted that in 1916 there was an outbreak of influenza at a British Army camp in Etaples, in northern France. Two months later an almost identical epidemic broke out at an army camp in Aldershot, England, headquarters of the British Army. One-quarter of the patients there died, and doctors noted similarities to the outbreak in France. Oxford pointed out that two years later, and within a short period of time, there were reports of flu outbreaks in countries that were very far away from one another. Norway, Spain, Britain, Senegal, Nigeria, South Africa, China, and Indonesia were all hit between September and November 1918. International air travel had yet to connect the world, so how did the virus spread so quickly? Oxford reasoned that it must have been "seeded" in these places much earlier, perhaps by demobilized soldiers returning home across Europe during the height of World War I in the winter of 1916.

Did the 1918 flu virus originate in France at the Etaples camp, or elsewhere, like Kansas? John Oxford pointed to photographs of French soldiers in contact with live pigs, chickens, and geese, but this doesn't prove that they were the source. Perhaps the origin was on the other side of the world, in China.

In June 1918 the *New York Times* reported that "a curious epidemic resembling influenza is sweeping over North China," with about 20,000 new cases reported. This epidemic predated the general outbreaks in Europe and America by a few months, and killed fewer people. The population seemed to have some immunity because of a previous exposure to a similar virus. Could a precursor to the 1918 flu have been circulating in China for several years before becoming a global pandemic? There was certainly a route for the virus to have spread from China to France. During the war, more than 140,000 Chinese laborers were recruited to France, and many were stationed near Montreuil—less than seven miles from the British Army's camp at Etaples. Mankind's dramatic movement around the globe was good news for an aspiring virus.

As the war in Europe entered its fourth year in 1918, many countries censored news reports, especially those containing information about a pandemic. There was enough bad news from the war without further depressing anxious citizens and soldiers. But Spain remained a neutral country throughout the war, so its press was free to report on the new influenza. This led people to think that Dr. Miner's "disease of undetermined nature" had started there. While scientists today continue to sort through origin theories, all agree on at least one point: what would be called "the Spanish influenza" certainly did not first erupt in Spain.

So where did the 1918 virus begin? Haskell County, France, or China? Knowing this might help prevent similar outbreaks in the future, but we still haven't figured it out. There is evidence to support each theory, but as the 1918 pandemic recedes into history, it is unlikely that we will ever come to a definite conclusion. This shiftiness, this uncertainty and mystery, are hallmarks of the flu's campaign against humanity.

Just as important as its origin and path were the details of its devastation. The world had yet to develop a treatment or discover antibiotics, and the consequences of influenza were extreme and unpredictable. How did this virus move, and what was it capable of? The answers to both questions are found on the bloody battlefields of Europe.

*   *   *

The virus attacked in two waves. The first began in the spring of 1918, as more than 110,000 U.S. troops deployed to the European front. It had been three and a half years since Britain and France had declared war on Germany and Austria-Hungary. Fighting now engulfed all of Europe. President Woodrow Wilson had declared in 1914 that the United States would follow a policy of "strict neutrality," but this became increasingly untenable as German submarines targeted American ships. Starting in 1917, the U.S. Army sent vast numbers of young men across the Atlantic into large, cramped camps that served as a perfect environment for the influenza virus. In the summer of 1918 the crowding became lethal. Influenza had mutated and young adults were particularly at risk. Soldiers lay within arm's reach of one another in vast sick wards, separated by nothing more than a hanging sheet.

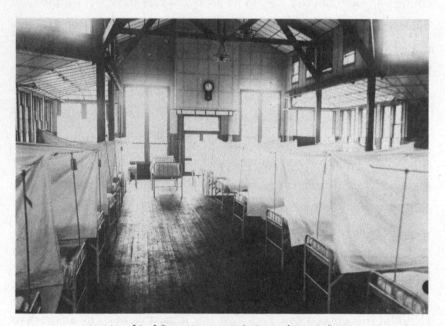

Interior of Red Cross House at U.S. General Hospital #16,
New Haven, Connecticut, c. 1918

This may explain why enlisted men died in much higher numbers than did civilians, despite an equal rate of infection. Most soldiers who were very sick were moved to these crowded wards, where they propagated the bacteria that caused fatal secondary infections. Instead of bringing patients back to health, these wards were large-scale petri dishes of disease.

The virus was not limited to the barracks and sick bays. Across Europe, tens of thousands of men were moving back and forth between home, army camps, the docks, and the war front. The U.S. War Department was sending 200,000 men a month to France. By the summer there were more than 1 million American soldiers fighting in Europe.

We do not know how many civilians became ill and died in that first wave. At that time there was no requirement for doctors to report anything about the flu. There were few state or local health departments, and those that existed were often poorly managed. However, we can get a sense of what happened by looking at statistics kept by the military. Starting in March 1918, there was a sudden increase in the number of influenza cases seen at Camp Funston in Kansas. Most of the troops recovered after two or three days of bed rest and aspirin, but 200 contracted pneumonia and around 60 died. In a sprawling camp of 42,000, these numbers did not make military physicians notice.

The situation in Europe was more drastic. One medical officer noted that influenza was raging through his division. His soldiers weren't able to march. By spring some 90 percent of the troops in the American 168th Infantry Regiment were sick with the flu. By June 1918 it had spread to French and British troops. There were more than 31,000 cases of influenza among British troops back in the UK, a sixfold increase over the previous month. On the European continent, over 200,000 British troops were unable to report for combat. The virus continued to travel, usually by sea. It struck Freetown in

Sierra Leone after a British steamship arrived in August with more than 200 of its crew enduring or recovering from influenza. In less than a week it had spread overland; before the end of September some two-thirds of the local population had become infected and 3 percent had died. Outbreaks were reported in Bombay. In Shanghai. In New Zealand.

This first wave was mild. Although many people became sick, the disease lasted only two or three days. Nearly everyone recovered. As usual, those who were most at risk were the very young and the elderly, whose mortality was much higher than the general population. But epidemiologists who examined death records noted that there was an increased incidence of death in the population between these two extremes. Young and middle-aged adults were dying from influenza at an unusually high rate.

When plotting influenza deaths against age, we most commonly see a U-shaped graph; one arm represents the very young and one the very old. In the middle, there are very few deaths. The same graph of the early 1918 deaths from the flu is shaped like a W. There still existed the high death rate at either end, but there was an additional spike that represented the young and middle-aged. Those most affected were in the twenty-one- to twenty-nine-year-old age range, a group usually considered to be the *least* likely to die from an infectious disease. This was peculiar and alarming.

Influenza and pneumonia mortality by age, United States. Influenza and pneumonia specific mortality by age, including an average of the interpandemic years 1911–1915 (dashed line), and the pandemic year 1918 (solid line). Specific death rate is per 100,000 of the population in each age division. *Institute of Medicine.* The Threat of Pandemic Influenza: Are We Ready? *Stacey L. Knobler, Alison Mack, Adel Mahmoud, and Stanley M. Lemon, eds. Washington, D.C.: National Academies Press, 2005, 74.*

By the time the first wave of flu hit the European continent it had virtually disappeared in the United States. In time, the numbers decreased in Europe too. By July 1918 the *British Medical Journal* reported that influenza was no longer a threat. But on both sides of the Atlantic the worst was still to come.

Perhaps the virus had mutated into a more lethal form. Perhaps the autumn brought people into closer proximity, where they were more likely to infect one another. Regardless, another wave of influenza began.

Among the earliest reports of the second wave was one from Camp Devens, some thirty miles west of Boston. The camp, built for about 36,000 troops, now housed more than 45,000. The outbreak began around September 8 and rapidly spread. Ninety patients a day came to the camp infirmary. Then 500 a day. Then 1,000 men each day, stricken with the flu. The infirmary was large, built to treat up to 1,200 patients. Soon it was not large enough. Eventually it housed 6,000 influenza victims. Bed after bed. Row after row.

"We eat it, live it, sleep it and dream it, to say nothing of breathing it 16 hours a day," wrote a young medical orderly in a letter dated September 29, 1918. He was assigned to a ward of 150 men, and his first name, Roy, is all we have to identify him. The grippe—another name for influenza—was all anyone could think about. One extralong barracks had been converted into a morgue, where dead soldiers dressed in their uniforms were laid out in double rows. Special trains were scheduled to remove the dead. For several days there were no coffins, and Roy wrote that the bodies piled up "something fierce." The orderly witnessed countless deaths, and he described what happened to the victims. Although it started as just another case of influenza, the infection rapidly developed into "the most vicious type of Pneumonia that has ever been seen." There were about 100 deaths each day at the camp. Among these were "outrageous" numbers of nurses and doctors. "It beats any sight they ever had in France after a battle," Roy wrote. He had witnessed the devastation and mayhem of the Great War, and still, there was no comparison. The flu was far worse.

Another eyewitness account of the carnage at Camp Devens was provided by Victor C. Vaughan, a prominent physician and dean of the medical school at the University of Michigan. In his memoir he wrote of ghastly pictures that hung in his mind, "which I would tear down and destroy were I able to do so, but this is beyond my power." One of those memories was of the division hospital at Camp Devens. "I see hundreds of young, stalwart men in the uniform of their country coming into the wards of the hospital in groups of ten or more," he wrote. "They are placed on the cots until every bed is full and yet others crowd in. The faces soon wear a bluish cast; a distressing cough brings up the blood stained sputum. In the morning the dead bodies are stacked about the morgue like cord wood." Vaughan was humbled by a plague he could not treat. "The deadly influenza," he concluded, "demonstrated the infe-riority of human interventions in the destruction of human life."

Less than a month after it began, the flu epidemic at Camp Devens had sickened 14,000 and left 750 dead. It swept across other military bases. Camp Dix in New Jersey. Camp Funston in Kansas. Camps in California and Georgia. At Camp Upton in New York, almost 500 sol-diers died. The flu was brought to Camp Dodge in Iowa by two service-men who arrived on September 12. Six weeks later more than 12,000 men in the camp had been infected. At one point the infirmary housed over 8,000 patients, four times its maximum capacity.

The outbreaks at each camp followed a pattern. First there were just a handful of cases, indistinguishable from a regular season of influenza. Within a couple of days the number grew exponentially, infecting hundreds, sometimes thousands. For three weeks the infirmaries were flooded and the death toll would rise. After five or six weeks the epidemic would disappear, as mysteriously as it had arrived. Some victims lingered with pneumonia, but there were no new cases, and life would slowly return to normal.

A great deal is known about flu in the army camps because of the record-keeping that the military required. But the second wave of flu struck beyond the military, killing tens of thousands in towns and cities across the United States. This wave is more challenging to piece together; nonetheless, by the time it subsided in the late spring of 1919 the death toll in the U.S. stood at 675,000 civilians and service members. The sheer amount of death is hard to fathom, and the spread of the disease nearly defies the imagination. Almost every town and city was affected.

In 1918 Philadelphia had a population of more than 1.7 million people. Like in most growing cities in the early twentieth century, its inhabitants were packed together in narrow tenement buildings. They were especially vulnerable to influenza because most of Philadelphia's doctors and nurses were overseas, tending to the wounded and the battle weary. As the flu struck, the few medical professionals left in town were stretched very thin. They were not prepared for what was to come.

Influenza probably snuck into Philadelphia in the middle of September 1918, around the time newspapers reported that the virus had made the leap from army camps to civilian neighborhoods. Spreading just as fast were rumors that German submarines loaded with germs were causing the outbreak. They weren't, but the Philadelphia Naval Shipyard likely was.

With a complement of 45,000 sailors, the yard had grown to become the largest naval base in the U.S. On September 7, 1918, the base welcomed 300 sailors who had been transferred from Boston. It is highly likely that a few of them were incubating the flu virus. Two weeks later, more than 900 sailors were sick. Officials stuck to a script: There was little to fear. The flu was nothing other than the usual seasonal germ masquerading under a new name.

But the virus was about to make a jump to civilians in a big way, and war bonds were partly to blame. Back in April 1918, a huge Liberty Bonds parade had taken place in New York City. The movie star Douglas Fairbanks addressed the masses who stood shoulder to shoulder. With his good looks and charming personality, he urged them to buy bonds to support the war effort. Five months later, it was time for Philadelphia to step up. The city planned its own war pageant to inaugurate the fourth Liberty Loan campaign on Saturday, September 28. It expected 3,000 fighting men, "and fighting women if need be," according to an article in the *Philadelphia Inquirer*. They would be joined by hundreds of factory workers and marshals who would keep the crowd singing together. All this was planned amid the city's flu epidemic. If there was any concern that such a large gathering would enable the flu to spread, it was overcome by the patriotic desire to participate.

The war-bond parade was essentially a marching band of influenza. The navy came down Broad Street as huge crowds looked on and cheered.

"It was a tremendously impressive pageant," declared the *Inquirer*, which estimated that more than 100,000 people thronged the streets. As people stretched their necks for a better view, they were also transmitting the flu virus person to person. The Liberty Bonds march had actually liberated the virus.

Only two days after the glorious parade, more than 100 people each day were dying from flu. In short order those numbers grew sixfold. Each day, health officials announced that the disease had passed, only to publish more grim statistics the next. The director of Philadelphia's Department of Public Health, Dr. William Krusen, issued an order to close schools, churches, and theaters. Perhaps if he had prohibited the Liberty Bonds march, the situation would not have been so dire. Placards reminded everyone that spitting on the streets was not allowed. They didn't do much good. In one day alone, sixty spitters were arrested.

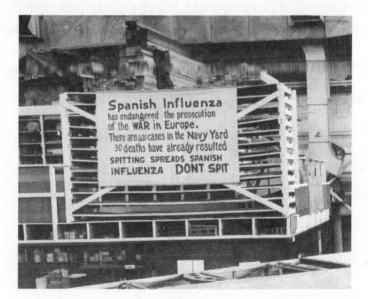

Naval Archive

With so many people taken ill, courts and municipal offices were closed and other essential services struggled without their employees. The police and fire departments were hobbled by diminished ranks. Because of severe staff shortages, Bell Telephone Company of Pennsylvania announced that it could handle only calls that were "compelled by the epidemic or by war necessity." With its regular hospitals over capacity, the city opened an emergency one. Within a day, all five hundred of its beds were full. Krusen called for calm and urged the public not to panic over exaggerated reports, but Philadelphia was being ravaged by what seemed like a biblical plague. Why *not* panic?

The only public morgue in the city had the capacity to hold thirty-six bodies. It soon held hundreds of corpses, most covered only with bloodstained sheets. For each available coffin, ten bodies lay waiting. The stench of death was everywhere. Local woodworkers dispensed with normal business and began making coffins full-time. Some funeral homes increased their charges by over 600 percent, prompting the city to cap premiums at "only" 20 percent.

The death toll in Philadelphia peaked around the middle of October and then, almost as suddenly as it had arrived, the plague subsided. Flu was still present, to be sure, but deaths from influenza fell back to their usual rates. The city slowly returned to a semblance of its former healthy self.

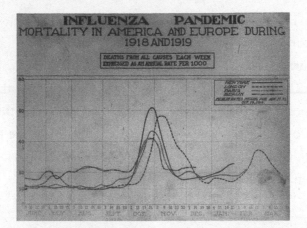

INFLUENZA PANDEMIC
MORTALITY IN AMERICA AND EUROPE DURING 1918 AND 1919

DEATHS FROM ALL CAUSES EACH WEEK
EXPRESSED AS AN ANNUAL RATE PER 1000

Historic chart of influenza death rates in New York, London, Paris, and Berlin,
1918–1919 *From the National Museum of Health and Medicine,
Armed Forces Institute of Pathology, Washington, D.C.*

What happened in Philadelphia was repeated across the United States, and the world. In San Francisco, the flu also peaked in October. More than 1,000 people died that month, almost double the usual number. The flu made it to Juneau, Alaska, which tried to prevent its spread by imposing a quarantine. The governor ordered that all disembarking passengers be examined by dockside physicians. Any person showing symptoms of the flu was refused entry to Juneau. This did not, however, prevent the entry of healthy-looking people who were carrying the virus but not yet showing symptoms. These carriers left Seattle and docked in Juneau a couple of days later, still within the flu's incubation period. When they arrived, they were briefly examined by a waiting physician, who, finding no signs of influenza, allowed them entry. This is the most likely way the virus snuck in. From there it spread to Native Americans in Nome and Barrow and dozens of remote villages. The devastation among tribes was even more profound than it was elsewhere. They were naturally isolated from the rest of the population and therefore lacked any antibodies to influenza. Half the population of Wales, a town of 300 on the west side of Alaska, was killed in the 1918 pandemic. In the tiny settlement of Brevig Mission, population 80, only 8 people survived.

This horror near the Arctic Circle would, in the long run, help fight the virus. The dead were buried in the frigid ground, and this resting place of permafrost preserved the bodies. Eighty years later, this allowed scientists to extract samples of the 1918 virus and, for the first time, identify its genetic code. For now, though, those bodies would lie in wait, frozen in earth and in time.

America was now fighting two wars. The first was against Germany and its military allies. The second was against the flu virus and its bacterial allies. In the words of one historian, it was a fight against germs and Germans.

As the Allies mounted a massive offensive on the western front, influenza attacked the ships transporting troops to the trenches of Europe. It killed many in the American Expeditionary Forces at the height of the Battle of Argonne Forest, in northeastern France. Just as the Great War entangled nearly every country in Europe, influenza ravaged the whole continent. At one French army base of 1,000 recruits, 688 were hospitalized and 49 died. Schools were closed in Paris, but not theaters or restaurants. Cafes stayed open as 4,000 Parisians died. Influenza leaped trench lines. German troops suffered too. "It was a grievous business having to listen every morning to the chiefs of staffs' recital of the number of influenza cases, and their complaints about the weakness of their troops if the English attacked again," wrote one German commander at the time.

In Britain, it was very much a "keep calm and carry on" approach. I was born and grew up in London, and even though I have now lived outside of Britain for most of my life, I recognize this reaction. Composure in the face of adversity and keeping a stiff upper lip were hallmarks of my childhood. I had seen such composure on the face of my grandmother as she recalled being evacuated from London during the Blitz, and I recognized it in the reaction to the Spanish flu a generation before. "Keep calm and carry on" were not just instructions for public behavior. They were part of the cultural DNA of the British themselves.

At first the newspapers barely mentioned the epidemic; when they did it was buried on the inside pages. The British government and a sympathetic press tacitly agreed to limit any discussion of the flu, lest it demoralize a public already weary of a world war entering its fourth year. The tension between reporting the facts and maintaining morale was embodied in a letter written by a Dr. J. McOscar that was tucked away in the back of the *British Medical Journal.*

"Are we not now going through enough dark days, with every man, woman, or child mourning over some relation?" he wrote. "Would it not be better if a little more prudence were shown in publishing such reports instead of banking up as many dark clouds as possible to upset our breakfasts? Some editors and correspondents seem to be badly needing a holiday, and the sooner they take it the better for the public moral [*sic*]."

Ironically, there was a detailed five-page report on influenza on the front page of the same issue in which this letter appeared. It underscored just how devastating the pandemic was. There had been a catastrophic outbreak among British and French troops, it noted, that had swept through entire brigades and left them unable to function.

Britain's chief medical officer also seemed reluctant to upset anyone's breakfast. His advice was limited: wear small face masks, eat well, and drink a half bottle of light wine. The Royal College of Physicians took a similar approach and announced that the virus was no more deadly than usual. The British seemed relatively unmoved throughout the saga. In December 1918, as the pandemic was ending, the *Times* of London commented that "never since the Black Death has such a plague swept over the face of the world; never, perhaps, has a plague been more stoically accepted."

Earlier that year, the medical correspondent for the *Times*, with what must have been a huge exaggeration, described a people who were "cheerfully anticipating" the arrival of the epidemic. The historian Mark Honigsbaum believes that this British stoicism was deliberately encouraged by the government, which had already worked to cultivate a disdain of the German military enemy. The same disdain was then directed at the influenza outbreak.

But whatever the attitude of the British toward the pandemic, influenza's toll was enormous. By the time it had subsided, more than a quarter of their population had been infected. Over 225,000 died. In India, then still a British territory, influenza was more lethal, with a mortality rate above the Empire's 10 percent. It was double that among Indian troops. In total, some 20 million Indians died as a result of the influenza pandemic.

Then there were Australia and New Zealand and Spain and Japan, and countries throughout Africa. All suffered, leaving us with that frightening, near-apocalyptic estimate: 50 million to 100 million deaths worldwide, in total. In the aftermath of this mass death—when the public was focused on "How?" and "How many?"—scientists were left to wonder: *Why?*

Was it the virus itself—perhaps a super version of the flu—or were there other reasons for its lethality? We've settled on four different explanations for why so many people died. Each is supported by some evidence, yet none is wholly satisfying.

The first explanation is that the virus had a protein on its surface that prevented the production of interferons, which signal to our immune system that our defenses have been penetrated. Healthy lung cells that transfer oxygen into the bloodstream are hijacked by the virus and destroyed by its replication process. Once dead, these cells are replaced with dull fibrous ones that are incapable of transporting oxygen, just like a scar that forms at the site of a cut and never looks the same as the surrounding healthy skin. An autopsy performed within hours on a U.S. Army private named Roscoe Vaughan in South

Carolina showed that one of his lungs had this type of pneumonia. It's possible that the sabotage of interferons allowed the 1918 virus to trigger a lethal viral pneumonia.

Second, if the 1918 virus itself didn't kill you, then secondary bacterial pneumonia probably did. Victims of the pandemic, their bodies weakened and their lungs already ravaged, caught bacterial infections like streptococcus and staphylococcus, which were deadly in this era before antibiotics. We now think that the majority of deaths in the 1918 pandemic resulted from these secondary infections, not from the flu virus itself. The South Carolina soldier's other lung showed evidence of this kind of infection. He was killed by the one-two punch of the virus and a bacterial infection that swept in as his defenses crumbled.

The third explanation for 1918's lethality is that the flu virus triggered an overreactive immune response that turned the body against itself. Suppose you cut your finger. Bacteria invade and infect the wound. Your finger becomes swollen, red, and warm because of increased blood flow to deliver more white blood cells to fight the bacteria. This inflammation, a painful but necessary development for fighting infection, is mediated by other kinds of messenger proteins called cytokines. Once the infection is overcome, cells stop producing cytokines and the immune system returns to its usual state of vigilance.

This return to normal didn't happen in many 1918 flu victims. Their lungs were hit by a "cytokine storm," an overproduction of these messenger proteins. In their exuberance, they began to destroy healthy cells along with invading ones. When a cytokine storm strikes, the immune response spirals out of control. The storm activates more immune cells, which release more cytokines, which activate more immune cells, and on and on. Large amounts of fluid pour out of the war-torn lungs. Healthy air sacs in the lungs scab over. It becomes harder and harder to breathe.

It's unclear why this storm occurred in some victims and not in others, or why it may have been especially common in those between twenty and forty years old. Infectious disease experts have called this the biggest unsolved mystery of the pandemic. If we solve it, we might be able to protect ourselves from another fatal plague of flu.

The fourth explanation points to the circumstances that surrounded the flu's transmission. It was a novel virus, having originated in birds. The virus then spent some time in another host, perhaps pigs or horses, until it emerged as a threat to humans. This occurred at a time when people were both penned in together—living in tenements or barracks—and unusually mobile, as the Great War circulated infected soldiers around Europe and beyond. Working-class families shared beds. Soldiers slept side by side in cots and traveled the world in steerage conditions. Had these human mixing bowls not existed, the flu virus, however lethal, would not have spread so quickly.

Today, influenza kills fewer than 0.1 percent of those who catch it. Nearly everyone recovers. In the 1918 pandemic most still recovered, but the death rate was twenty-five times greater. So many died in the U.S. that the average life expectancy in 1918 fell from fifty-one to thirty-nine years.

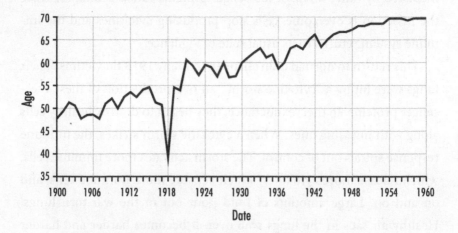

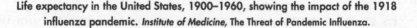

Life expectancy in the United States, 1900–1960, showing the impact of the 1918 influenza pandemic. *Institute of Medicine,* The Threat of Pandemic Influenza.

*   *   *

In December 1918, in the midst of the pandemic, 1,000 public health officials gathered in Chicago to discuss the plague, which had killed an estimated 400,000 people over a three-month period. Some were already predicting that the following year there would be an even more virulent flu outbreak.

Dr. George Price, one of the attendees, described the state of affairs in his report. It makes for terrifying reading.

First, the doctors admitted that they didn't know the cause of the pandemic. "We may as well admit it and call it the 'x' germ," Price wrote, "for want of a better name." Physicians had identified several distinct microorganisms in the secretions of the victims, but were they the cause or were they opportunistic hijackers of a body already beset by disease? (The latter, it turned out.)

Attendees at this conference did agree on a few things. Whatever transmitted the disease was found in the spray and mucus from the throat, nose, and mouth. It could be spread through droplet infection by sneezing and coughing, and by hand-to-mouth contact. This prompted one physician to suggest that the only way to reduce the spread was to put "each diseased person in a diver's suit."

Doctors also agreed that if you recovered from this influenza, you emerged with a certain degree of immunity. Many people over forty years old were spared. The theory then, as now, was that this demographic—those who had lived through a severe influenza epidemic in 1898—had acquired immunity against the 1918 infection.

But how to control the disease? The conference erupted into heated discussions inflamed by a general despair. The flu had spread despite precautions to combat it, and then it had suddenly and unexpectedly disappeared. Face masks, which were then being worn by a large number of the general public, were no guarantee of protection. Many health officials believed they provided a false sense of security. Perhaps that was true, but there was still a value in providing any kind of security. Chicago's health commissioner made this clear. "It is our duty," he said, "to keep the people from fear. Worry kills more people than the epidemic. For my part, let them wear a rabbit's foot on a gold watch chain if they want it, and if it will help them to get rid of the physiological action of fear."

Officials tried to collect data on the sick and the dead, but many states were still not required to report their cases. Physicians on the front lines of the disease were too busy to fill out the necessary paperwork, and plenty of victims died before they ever came in contact with the medical system. It was thus nearly impossible to estimate the number who had died, or had been infected and recovered. The virus snatched people before they could even be counted. No mechanisms existed to render the monstrous plague in practical, numerical terms.

During the bubonic plague of London in the 1600s, many afflicted households painted a large cross on their front door, along with the words "Lord have mercy on this house." It warned that illness and death lurked inside. Something similar happened in 1918—but in a more regimented way—with "placarding," the posting of signs on front doors. Placarding was supposed to warn healthy people to stay away, but in many communities almost every household door was marked.

There was also a public health effort to reduce crowding and mixing in public spaces by closing schools, theaters, and stores. It was a way to force people to spend their leisure time resting, to store up energy and stave off infection. But it was not clear whether these closures actually helped. Detroit shuttered very few public spaces and suffered only a relatively minor outbreak, while Philadelphia instituted a much tougher closure policy that didn't prevent a health catastrophe. Royal Copeland, the president of the New York Board of Health, changed the schedule of buses and subways to combat overcrowding. He installed large signs around the city that reminded the public not to spit. But he did not close schools or theaters; he argued that because many schoolchildren lived in crowded tenements they would be better off in school, where they could be taught how to stay healthy.

Dr. Price's description of the 1918 conference in Chicago ended with a clarion call to action. Despite great uncertainty and a degree of despair, he maintained that the best way to end the influenza epidemic was through public health policy. There needed to be better coordination between the health agencies, which should be placed under a unified command, like the military. To beat the enemy, private and community institutions needed to work together, at the municipal, state, and federal levels. Price knew he was asking for the moon. The virus required nothing less. Among the symptoms of influenza was something more pernicious than a fever or shortness of breath. It was a feeling of futility, which had a lifelong impact on Victor C. Vaughan, that dean of the medical school at the University of Michigan. Having witnessed the deaths of so many, Vaughan decided "never again to prate about the great achievements of medical science and to humbly admit our dense ignorance in this case."

The history of the 1918 influenza pandemic is depressing reading. It's like watching a horror movie that you have seen before. You know who the killer is, but you can't jump in and save the victim. However, during the pandemic and in the years that followed, there was a steady stream of dramatic medical discoveries that would, for the first time, allow us to fight back against the flu.

Some medical professionals were so desperate to identify what it was that caused influenza that they placed their own lives in peril. In the winter of 1918–1919, at the height of the flu epidemic, about 30 million Japanese became ill and more than 170,000 of them died. Despite this, a professor named T. Yamanouchi was able to find fifty-two doctors and nurses who offered to help by becoming human guinea pigs. The professor took "an emulsion of the sputa" from influenza patients and placed it in the noses and throats of this group of volunteers. Some received this contaminated goop directly, while others got it after it was run through a filter whose mesh was fine enough to capture all bacteria. Both groups soon exhibited signs of influenza, which led the Japanese researchers to affirm that no known bacteria could be the cause of influenza. In addition, they concluded that the disease could be spread by getting into the nose or throat of its victim, a feature of flu that we now take for granted but was barely recognized at the time.

There have always been researchers willing to experiment on themselves. The Australian physician Barry Marshall is a recent example. He co-discovered the bacteria that cause stomach ulcers. In order to prove this, Marshall himself agreed to drink a sludge containing the bacteria and see what happened. He got stomach ulcers. And a Nobel Prize. But the courage of those Japanese volunteers in 1918 was even more remarkable. All around them an epidemic was killing its victims in unprecedented numbers, and there was no known cause or cure. Yet fifty-two doctors and nurses agreed to be inoculated with material from those who had been infected. They were prepared to make the ultimate sacrifice. Their bravery and generosity almost defies belief.

The Japanese discovery was quickly replicated. In 1920, two American researchers also developed a filter small enough to remove all known bacteria from nasal washings taken from those with influenza. Yet the remaining material was still able to cause influenza-like symptoms when given to live rabbits. Again, the conclusion was that bacteria could not be the cause of influenza. Soon there were reports of other illnesses that were caused by agents too small to be stopped by a bacteria-catching filter. The cause of the pandemic was still a mystery, but we had eliminated bacteria as a suspect.

So what was getting through those bacterial filters? It was, of course, the influenza virus. In 1933 two British scientists—working at a laboratory in north London only a few miles from where I grew up—demonstrated that they could infect ferrets with samples obtained from the throats of patients that had been filtered to remove all bacteria. (It turns out that ferrets are one of the few mammals that get sick with influenza. They are also easier to work with than pigs.) This built on the work of the Japanese group, and the British scientists concluded that "epidemic influenza in man is caused primarily by a virus infection." Another important advance in the same decade was the discovery that the influenza virus could be cultivated. It was injected into the amniotic fluid of developing chicken embryos, which turned out to be a perfect growing medium for the rather picky virus. This was an incredibly important development. If you can grow the virus, you can also collect it, kill it, and inject it into healthy people. And then you have a vaccine.

Finally, in 1939, there was a watershed moment in the history of virology. The newly invented electron microscope took a picture of a virus. For the first time in history we could actually see the culprit. By the 1940s, scientists had isolated two strains of influenza (A and B) and begun to test vaccines; one of these scientists was Jonas Salk, who later developed the polio vaccine. After Crick and Watson's discovery of DNA in 1953, it didn't take long to identify the various building blocks of a virus. The field of virology could then develop tools and techniques to identify viruses and classify them based on their genetic components.

Medicine is the art of diagnosing, treating, and curing. It is also the art of preventing history from repeating. Did we learn enough from the 1918 pandemic? Could the lessons learned prevent another catastrophe? We now knew what virus we were up against, but could we do a better job of fighting it? The world would be tested again a few decades later, when the next pandemic arrived on the island of Hong Kong.

4

"AM I GONNA DIE?":
ROUND TWO, AND THREE, AND FOUR . . .

Thousands of sick people stood in line, awaiting treatment in clinics. On their backs, women carried glassy-eyed children, who in turn carried the virus. Officials called it the "worst epidemic outbreak in years." It was April 1957, half a lifetime after the 1918 flu, and at least 10 percent of Hong Kong's 2.5 million people were sick. There was no end in sight—especially as an influx of 700,000 refugees from communist China congregated in overcrowded neighborhoods.

The world appeared on the brink of the first global influenza pandemic since 1918, although this outbreak was caused by a different strain of the flu virus. It was labeled H2N2, and contained the genes of both a human and an avian flu virus. The avian genes had likely leaped from ducks, which were (and continue to be) handled in great numbers in the bird markets of China. It was a textbook example of antigenic shift, which created a new virus not recognizable by our immune systems. The results could have been catastrophic, but the virus seemed capricious. True to its unpredictable nature, the flu hit people differently. Identical samples were found both in those who had died and in those who suffered mild symptoms and recovered.

This "Asian influenza" invaded Britain by the autumn, tripling the flu death rate. It killed about 68,000 people in the United States and perhaps 2 million worldwide. However, unlike the 1918 pandemic, the Asian flu did not overwhelm those who were otherwise healthy. Instead, it appeared to target people who were always at risk, like those with chronic underlying heart or lung diseases. In the U.S., Asian flu also quickly spread to schoolchildren; more than 60 percent became clinically ill.

The story of the twentieth century can be told through its wars or its technological leaps—or through its flu pandemics. The outbreaks were irregular in timing, as with all of recorded flu history, but they were familiar in progression: a point of origin, a rapid spread, sickness and death, and a furious public debate about how to respond. With each passing decade, from the Eisenhower years to the Gerald Ford interlude to the Obama era, we became more equipped to fight back, but the counterattack was never perfect and always fraught.

Medically speaking, 1957 was very different from 1918. Physicians had two new weapons with which to fight the pandemic. The first was antibiotics, which were now available to fight any secondary bacterial infections. They changed everything. Deadly skin infections were easily treated. Rheumatic heart disease, a dreaded complication of strep throat, became a thing of the past. Perhaps most important of all, bacterial pneumonia could now be treated. In the pre-antibiotic era, pneumonia was called "the old man's friend" because it caused a very rapid but blessedly painless death. Breathing became more and more difficult as the bacteria multiplied within the lungs, preventing the entry of life-sustaining oxygen. There was nothing to be done. The patient would drift off into unconsciousness and soon expire. These lethal bacterial pneumonias often followed an influenza infection, especially in the elderly. And now there were, at last, antibiotics that could save thousands of lives—which is exactly what they did in the 1957 influenza outbreak. Without them, the death toll would have been far greater.

The second new weapon was not a cure for those who had been stricken, but a prevention for those who were still healthy. For the first time there were vaccines that could protect those who were not yet infected. The American virologist Maurice Hilleman first learned of the 1957 outbreak from an article in the *New York Times*, which prompted him to get a head start on the production of an H2N2 vaccine. Working together with the pharmaceutical industry, he spent much of the summer of 1957 creating a vaccine that could thwart Asian flu. Growing enough influenza virus to produce the vaccine is challenging. For reasons that are not known, the virus is very picky about what it chooses to grow on. After a long process of trial and error, it became apparent that the only medium on which influenza would grow was fertilized chicken eggs. And so Hilleman went to work, and demanded that farmers keep enough chickens available to produce millions of eggs. Thanks to his leadership, there were about 40 million doses of the vaccine by the end of 1957. He recalled later, "That's the only time we have ever averted a pandemic with a vaccine."

The H2N2 influenza continued to circulate for the next decade, but it no longer caused significant outbreaks. Eventually it disappeared, only to be replaced by a new strain of avian flu, which in 1968 caused the third pandemic of the twentieth century. It originated in Hong Kong (again!), and like the 1918 influenza, its spread was accelerated by war.

The "Hong Kong influenza" was a descendant of the 1957 Asian flu. Hong Kong remained ground zero because chickens and ducks were handled in vast open markets, once again allowing the avian and human influenza viruses to mix and swap genes. It quickly spread to Southeast Asia, and in August 1968 soldiers returning from Vietnam brought the virus into the United States. It was reported in Australia and England by September, in Canada by December, and in France in January of 1969.

The new flu virus replaced the H2 protein on its surface with a different one, called H3, but the rest of the virus was pretty much unchanged. Because of these similarities, the old vaccine to the Asian H2N2 influenza provided some immunity against the Hong Kong flu. Similarly, those who had caught the 1957 Asian flu retained a degree of immunity to the new infection. This explains why the Hong Kong flu, which caused a million deaths worldwide, was still less deadly than the Asian flu. The Hong Kong flu strain still causes influenza today, although not on an epidemic scale. Our immune system has learned to fight this flu virus as it would any other.

The twentieth century had witnessed the mass death caused by the Spanish flu, and was now facing new influenza viruses. The virus was never wholly vanquished. It was always changing shape. And it next struck almost two decades later, when an army private collapsed at Fort Dix, outside of Trenton, New Jersey.

When new recruits arrived at Fort Dix in the fall of 1975 they were given the most current flu vaccination, which contained weakened versions of common flu strains. After the Christmas break, soldiers returned to base during an especially cold winter. In short order there was an outbreak of a flu-like illness causing fevers, sore throats, and fatigue. Throat swabs of the patients showed not influenza but another microbe, adenovirus, which causes an influenza-like illness. The patients were also tested for the flu virus by the New Jersey Department of Health, and the results were puzzling. Some men had a known strain of influenza, but several had a mystery strain. Two of these mystery swabs were sent to the Center for Disease Control, which identified the virus as a descendant of the 1918 virus.

All the recruits recovered, but then, on February 4, 1976, a private named David Lewis collapsed after a routine five-mile march and was taken to the Fort Dix hospital. There, a few hours later, he died from what appeared to be a fast-acting pneumonia. Lewis's death was initially reported as having been caused by a disease of unknown origin. But within two weeks the CDC identified the virus: it was a strain of pig influenza.

This prompted more questions than answers. None of the men who were sick at Fort Dix had been in contact with swine. The virus must have mutated, allowing it to spread directly between people, without going through its original host species. The CDC also identified the virus as an H1N1 type—like the one in 1918.

The Fort Dix outbreak was caused by a new strain of influenza, but it was most certainly *not* an epidemic. Only one person died. Nonetheless, it was of pivotal importance. To this day, the way the government responded to the 1976 outbreak is a topic of considerable controversy. Some experts believed then that the potential for a pandemic demanded a massive vaccination program. Others said the risks of such a program outweighed the benefits.

Because the strain originated from a virus that infected pigs, that 1976 flu was called swine flu, and it became a public health emergency. No one knew if it would spread like its 1918 ancestor and become a pandemic, or remain a local outbreak that caused only a single death. In the midst of the initial effort to identify the Fort Dix virus, Edwin Kilbourne, a virologist at the Mount Sinai School of Medicine in New York, published an op-ed piece in the *New York Times* with a gripping headline: "Flu to the Starboard! Man the Harpoons!" Kilbourne was not aware of the Fort Dix outbreak when he wrote the piece, which makes his observations all the more poignant.

Kilbourne looked back at the frequency of previous flu pandemics and discovered that they occurred at intervals of eleven years or less. The next pandemic, he wrote, was due sometime before 1979. It could be minimized by vaccinating the 45 million Americans who were most at risk. He also called on the CDC, the Food and Drug Administration, and the National Institutes of Health to work together to provide flu vaccines for the public, and called on health officials to plan for "an imminent natural disaster."

The day after Kilbourne's prescient publication, federal officials gathered in Atlanta at the CDC headquarters. How should they respond to the Fort Dix swine flu? At a quiet and understated press briefing, they released some details of the outbreak, though they were careful never to mention the 1918 pandemic. Swine flu remained confined to Fort Dix and had not spread to the surrounding civilian population. But officials were concerned that the virus would reappear the following fall and spark a global pandemic. The only way to prevent this was to quickly produce a vaccine, and they decided that the best scientist to lead the project was Kilbourne himself.

In March, less than a month after Private Lewis died, officials decided to turbo-charge the production of a vaccine and inoculate *all* Americans by the autumn. While the risk of a pandemic was very small, the consequences would have been devastating. "Better a vaccine without an epidemic," Kilbourne said, "than an epidemic without a vaccine."

A massive vaccination program was also a massive gamble. If no swine flu outbreaks occurred in the fall, health officials could be accused of waste and overreach. But there was another concern: in a letter to the *New York Times*, Dr. Hans Neumann from the New Haven Department of Health noted that based on the projected scale of the immunizations, within two days of getting a flu shot, about 2,300 people would have a stroke and 7,000 would have a heart attack. "Why?" he asked. "Because that is the number statistically expected, flu shots or no flu shots."

Likewise, in the week following a flu vaccine, another 9,000 people would contract pneumonia, of whom 900 would die. These would certainly occur after a flu shot, but not as a *consequence* of it.

"Yet," wrote Neumann, "can one expect a person who received a flu shot at noon and who that same night had a stroke not to associate somehow the two in his mind?"

Grandma got the flu vaccine in the morning, and she was dead in the afternoon. Although association does not equal causation, this thinking could lead to a public backlash against vaccinations that would threaten future programs. (More recently we have seen the mistaken identification of association with causation produce a backlash against vaccinations, which are mistakenly blamed for autism.) As a preventive measure, some health experts advised against the vaccine, and said it should be stockpiled instead. If the swine flu returned in the fall, then—and only then—would the vaccine be released to the public.

The decision made it all the way to the White House. President Gerald Ford accepted the advice of his health care advisers, who wanted to err on the side of action rather than inaction. Flanked by Jonas Salk and Albert Sabin, the discoverers of the polio vaccine, Ford announced a request for emergency funds to inoculate every man, woman, and child against the impending threat of pandemic swine flu. Discarding the caution of doctors at the meeting in Atlanta, the president mentioned the 1918 pandemic.

"Some older Americans today will remember that 548,000 people died in this country during that tragic period," Ford said in the White House briefing room. "Let me state clearly at this time: No one knows exactly how serious this threat could be. Nevertheless, we cannot afford to take a chance with the health of our nation."

His decision would have far-reaching implications. There were complexities for industry too. Drug companies faced the challenge of producing enough vaccines in short order, and of obtaining insurance in case anything should go wrong. After the manufacturers threatened to cease all production of the vaccine unless they were adequately protected, President Ford signed a bill in August to indemnify them. The vaccine program started in early October, and was quickly followed by precisely the kind of frightening media reports that officials had feared. It became a public relations nightmare.

Three elderly people died after receiving the flu vaccination at the same clinic, and people panicked about a connection that didn't exist. Walter Cronkite appeared on the evening news to reassure the public, and to caution against sensationalist reporting. His pleas went unheeded. The media blamed the vaccine for all manner of illnesses and deaths. The *New York Post* even claimed that it had been used as a deadly weapon to kill the head of the Gambino crime family. The CDC reminded the public that in truth there was no increase in the death rate among elderly Americans receiving the vaccine. President Ford got his flu shot on television. But public opinion was swayed not by logic or evidence but by emotion and anxiety.

People doubted the vaccine or, worse still, feared it. There were reports of an increase in the number of cases of a rare neurological disease called Guillain-Barré syndrome (GBS), which causes a range of symptoms including difficulty swallowing, weakness in the arms and legs, and muscle paralysis. In the fall the CDC recorded an unusual number of GBS cases following vaccination. Although there was no known relationship between flu vaccines and GBS, the CDC asked doctors to report any new cases to them. This added to the uproar. Some doctors who couldn't diagnose the source of a patient's weakness were now attributing it to GBS, especially if the patient had recently been vaccinated. By December the CDC was so concerned about the confusion that it ended the vaccination program. There had not been a single case of swine flu, but there were dozens of cases of GBS attributed to the flu vaccine. A "sorry debacle," the *New York Times* wrote in an editorial. The newspaper faulted "the excessive confidence of the Government medical bureaucracy" led by the CDC, which had taken advantage of the outbreak to multiply its budget. *Newsweek* was more direct. This was "the swine flu snafu," and David Sencer, the CDC's director, was forced to resign.

And then the lawsuits began (after all, this is America). Vaccine manufacturers had been indemnified by Congress, which left the federal government liable for any damages. By 1980 more than 3,900 claims had been filed, seeking a total of at least $3.5 billion in compensation. By then over 500 cases of GBS following a flu shot had been reported, and 23 of those people had died.

Despite more than forty years of analysis and debate, it is still unclear if there really was a link between GBS and the swine flu vaccine. In the military, where almost 2 million people received a double dose of the influenza vaccine, there was actually a *decrease* in cases of GBS. Today, on its seasonal influenza website, the CDC states that data on the association between GBS and seasonal flu vaccination are "variable and inconsistent" across flu seasons. But even if there had been a causal association, it was, for some, a small price to pay for avoiding catastrophe. Thirty years later, Sencer, the fired CDC director, reflected on the decision. "Public health leaders must be willing to take risks on behalf of the public," he wrote. And even with 20/20 hindsight, Sencer stood by the decision to release the vaccine, because "when lives are at stake, it is better to err on the side of overreaction than underreaction."

Edwin Kilbourne, the virologist who had written that prescient op-ed piece in the first days of the 1976 outbreak, also defended his decisions. He declared an "unyielding position on the need for vaccine production and immediate vaccination." Even though the outbreak was confined to Fort Dix, the swine flu was transmissible from person to person and was in the same family as the 1918 virus. That the virus disappeared over the summer of 1976 meant nothing. Viruses can vanish, only to return as an even greater threat. That had happened after the first wave of the 1918 epidemic. It was likely to happen again. "None of these facts," he wrote, "was noted by critics of the program." Kilbourne called for better influenza preparation, "but with the realization that no amount of hand washing, hand wringing, public education, or gauze masks will do the trick."

The impulse to do *something*, to react in the face of a catastrophe, is a common theme in our fight against influenza. In 1918 Chicago's public health commissioner noted that "worry kills more people than the epidemic," and so every step should be taken to assuage the public. In 1976 we stockpiled vaccines for every single American, at great cost, even though there was no certainty that a pandemic was coming. It was a high price to keep the public from worrying.

The next "pandemic" broke out in 2009, in the era of social media and the twenty-four-hour news cycle, and it was also caused by a swine flu strain. It invaded a world that had already endured, minute by minute, the traumas of the 9/11 terrorist attacks, the Indian Ocean tsunami, and Hurricane Katrina. But now there were regular planning exercises and heightened cooperation between federal, state, and local health agencies. Life was different. And there was Twitter.

In March 2009 the virus was first detected in Mexico, where it killed about sixty people. Acting quickly, the Mexican government closed schools, banned public gatherings, and ordered troops to hand out face masks at subway stations. By April the strain had reached the United States, and some students in New York tested positive for the disease. This time the strain contained genes from four ancestors: an American swine flu, a European swine flu, an avian flu, and a human flu. It was still an H1N1 strain, similar to the 1918 and 1976 strains. By June there were more than 30,000 cases across seventy-four countries, and the director-general of the World Health Organization (WHO) declared it a pandemic. Over half the cases of influenza in the United States were caused by the new strain. While most of the deaths occurred in children and adults, there were very few in those over the age of sixty-five, who appeared to be immune; perhaps they had been infected by a similar virus years prior. By June, all fifty states had reported cases of H1N1, and the CDC reported that at least a million people had been infected. Thankfully nearly all of them recovered without needing medical treatment.

Once again, the latest flu outbreak bore eerie similarities to the 1918 pandemic. Cases were first reported in late spring and early summer. Next, the virus went into hiding. Then, at the end of August, just like in 1918, there was a sudden surge in cases. But for the first time in history, there were now drugs that could target the influenza virus itself. These medications were available by prescription to the public, and were also part of the country's Strategic National Stockpile (SNS), a mother lode of medications and equipment that could be released in a medical emergency that overwhelmed the health care system. To meet the demand in 2009, the SNS released some of its store of antiviral medications as well as almost 60 million face masks.

The FDA also authorized the release of an experimental drug called peramivir. It was part of the SNS but was still undergoing clinical trials, and there was very limited data on its safety and efficacy. As a result, it could be used only under extraordinary conditions, and the 2009 outbreak qualified. The FDA received 1,371 requests for the drug between October 2009 and June 2010. After the pandemic subsided, doctors went back to look at the success of peramivir, but were unable to draw any definitive conclusions. About 15 percent of the patients who received it died, but they were already critically ill when the request for the drug was made. Three years later the FDA approved peramivir, even though there was little evidence of its healing magic.

Over the summer of 2009, vaccine production kicked into gear. Following Gerald Ford's example, President Barack Obama was photographed that December rolling up the sleeve of his sweater in the White House to receive his influenza vaccine. Again, a U.S. president was assuring the public, through the media, that the vaccine was both safe and necessary.

"People need to understand that this vaccine is safe," Obama said during a radio interview in the Oval Office, before noting the low rate of vaccination among African Americans. "If I had the two people that are most important in my life, my two daughters, get it right away—and they've been just fine with it and in fact haven't gotten sick this entire flu season—then you need to know that you need to make sure your children get it as well."

But the number of cases had already peaked back in October. By January the flu season returned to its baseline. The swine flu crisis fizzled out. Experts had predicted it would cause up to 1.9 million deaths in the United States, but the estimated toll was 12,500, an unusually low figure for an influenza outbreak. Worldwide, deaths from the pandemic also did not exceed the number during a usual flu season.

The most lasting side effect of the 2009 swine flu pandemic was confusion among the public. Officials warned us repeatedly to brace for a deadly winter. The media reported flu deaths and advice on how to avoid getting sick. In an interview with the *Washington Post*, a fourteen-year-old from Maryland described her fears as she developed a high fever. "I went to the doctor when it hit 103," she said. "He did a rapid test. He said he was pretty sure it was swine flu. It was not the regular flu season, and he had six cases that day. . . . When he said that I had swine flu, my mom and I laughed. 'Okay, what do I *really* have?' He said, 'Yeah, I think that's what you have.' Oh, my God. At first I thought, 'Am I gonna die?'"

Her fears were not surprising, given the mixed messaging from the top. In April 2009 President Obama had said that there was no cause for alarm. Then in October he declared the H1N1 outbreak to be a national emergency. The public didn't know what to think.

To complicate matters, there was another kind of viral outbreak, on Twitter and other social media. For the first time in a public health emergency, misinformation and fear were spreading faster than the virus. There were almost 3 million tweets about the flu. It now had its own hashtag: #swineflu. A CDC spokesman thought that the online buzz about swine flu was a good sign, showing that the public was engaged and ready to fight back with knowledge and prevention measures. But Twitter feeds were infected with fearmongering, and cable news outlets like CNN and Fox News were criticized for exaggerating the story and stoking public worry. If the public believed that scientists were "crying wolf," they would disregard future warnings.

In Washington, D.C., my emergency department was inundated with incoming patients who had flu-like symptoms. The ones with the flu were easy to spot. They wore blue disposable face masks handed out by the nurse when they checked in. Had we tested them for it, many would have been positive for H1N1. But it didn't matter whether they were ill with swine flu or seasonal flu or just another viral infection, because almost all were well enough to be discharged. The 2009 swine flu season turned out to be no more and no less deadly than any other flu season.

Then came the fallout. There were claims that the death estimates were exaggerated. Fiona Godlee, editor of the influential *British Medical Journal*, reported that some of the experts advising the World Health Organization had not disclosed their financial ties to the pharmaceutical industry. This raised ethical questions about conflicts of interest.

The real problem was the WHO's use of the word "pandemic." Most people think of a pandemic as a disease that spreads and kills thousands of people. That description is echoed in the WHO's official definition of the word as an infectious disease that causes "enormous numbers of deaths and illness." But in talking about the 2009 outbreak, the WHO used a more academic and narrow definition that focused only on prevalence, not severity. After this was pointed out by an astute CNN reporter, a WHO spokeswoman announced that the organization had erred in using the more apocalyptic definition. "It was a mistake, and we apologize for the confusion," she said, noting that the word painted "a rather bleak picture and could be very scary."

The H1N1 outbreak was just another kind of seasonal influenza, and no more dangerous than usual. One word had elevated its stature in the minds of the media and the public. The 2009 "pandemic," which was not really a pandemic at all, taught us that language is both a weapon and a handicap when waging a campaign against influenza. The public took the WHO and the CDC at their words, imagining that a lethal 1918-style outbreak was about to be unleashed.

This was not the first time that the description of a pandemic did not match its severity. The 1957 Asian flu was confusingly described by the WHO as both "comparatively mild" and "substantial." The 1968 Hong Kong flu outbreak was called "mild" by the WHO, and "moderate" by the CDC. Either way, the influenza historian John Barry noted that few people who lived through the 1968 pandemic "even knew that it occurred."

After the devastation of 1918, the flu spent the rest of the twentieth century stalking us without delivering another gut punch to the global population. But as our knowledge of influenza accumulated, the virus found new weaknesses in our systems. It revealed inadequacies in policy, preparedness, response, and media reaction. And we still didn't have a genetic profile of the 1918 virus itself. But that was about to change. The hunt for a sample of the original 1918 virus involved a medical student working in the Arctic, a young pathologist trying to save his job from Congress, and bodies dug up from the frozen tundra.

5

RESURRECTING THE FLU

The National Institutes of Health in Maryland keeps samples of the 1918 flu virus in a freezer at an undisclosed location. It's not easy to get anywhere near that locked freezer, let alone inside it. First, you have to get onto the campus of the NIH, which requires identification, a reason to be admitted, and a PhD, preferably in one of the life sciences. Once you get through and find the building, a guard has to buzz you in via an airlocked entrance with double doors. Inside, you will pass through a metal detector and then be firmly guided toward a locker, where your cell phone, thumb drive, computer, pager, and camera must be deposited. Then, and only then, will you be escorted farther into the building.

Jeff Taubenberger makes this trip every day. He is the chief of the Viral Pathogenesis and Evolution Section, a laboratory within NIH. His unit houses a couple of dozen scientists, postdoctoral students, and fellows researching the influenza virus, who refer to it simply as "1918." Their offices sit around the perimeter of a rectangle of sealed labs. In one of those labs, in a freezer, rests the 1918 virus, frozen and tamed. It took an epic effort to bring 1918 back from the dead. Scientists traveled to the ends of the earth, scavenging for stowaway viruses in buried corpses. Researchers hunted through dusty archives and painstakingly reconstructed genomes. If 1918 was purely a thing of the past, we couldn't study it properly. We had to keep 1918 present. It was a difficult and dangerous proposition, and it started with a gimmick.

After completing medical school, Taubenberger began his career at the NIH, where he trained as a pathologist. In 1993, not long after getting his PhD for work that focused on stem cells and lymphoma, he was recruited by the Armed Forces Institute of Pathology (AFIP) at the Walter Reed Army Medical Center, a few miles away. There he was to set up a new department of molecular pathology, which would demystify diseases using DNA analysis. In the early 1990s new lab tricks and techniques allowed pathologists to analyze the DNA of tissues that had been biopsied and embedded in small paraffin cubes. This was a big deal because until then, it was only possible to analyze the DNA from frozen specimens, which involved costs and complications. In contrast, paraffin-embedded samples can be kept on a shelf in the lab. Taubenberger studied ways to handle these tissues, and hadn't given a thought to flu. But then Congress got involved.

In 1994, with majorities in both the House and Senate, Republicans became locked in a series of nasty partisan battles with Democratic president Bill Clinton. In one of the many skirmishes over spending cuts, Congress toyed with the idea of abolishing the AFIP, where Taubenberger had recently been appointed the department head. As a result, he was under pressure to show Congress that the institute was worth keeping.

One way to do that was to demonstrate the scientific value of the tissue samples it contained. Taubenberger knew that the records of all its specimens were computerized and therefore searchable—all of them— going back nearly a hundred years. Perhaps, he thought, the institute might house an original tissue sample from a victim of the 1918 pandemic. If it did, the new techniques would allow him to sequence the genetic code of the virus. Now, that would be impressive—and surely enough to demonstrate the value of the institute in an era of cost-cutting.

He combed through specimens, using terms like "flu," of course, and "*gripe*," the Spanish word for "flu." He found twenty-eight samples. Now he could apply the techniques of his molecular pathology lab. Usually, he would work to identify the genetic characteristics of a cancer in a living patient, which could then help doctors to identify a targeted therapy. This time, however, he wanted to reveal the genetic building blocks of a virus long dead.

To begin the process of uncovering the genetic code of 1918, Taubenberger needed to find the right *kind* of specimen. Like all flu viruses, 1918 reaches peak replication two days after infection. After about six days the virus stops multiplying, and it is no longer found in the lungs. This meant that tissue from a patient who had contracted the 1918 virus and died from a bacterial pneumonia several days later could not be used. Their tissue would not contain any viral particles; instead, it would be overrun with the bacteria that so often followed the viral infection.

So Taubenberger and his team had to find samples from patients who had died within a week of the initial symptoms. In one sample, tissue from each of the victim's two lungs showed slightly *different* pathological changes. In one lung researchers found bacterial pneumonia, which was useless in this endeavor. But the other lung showed acute swelling of the walls of the tiny bronchi. This was enormously significant because this swelling is seen only in acute *viral* pneumonia, which meant Taubenberger had discovered a pathologist's smoking gun: he knew that while most of the victims had died from the *complications* of the influenza virus, this victim had certainly died from lung damage caused directly by the virus. He named this victim "1918 case 1," and this sample would be the key to identifying the genome of 1918. The sample belonged to Private Roscoe Vaughan.

On September 19, 1918, at Camp Jackson near Columbia, South Carolina, Private Vaughan came down with a fever and chills. One week later he died. After an autopsy, small specimens of his lung were preserved and set into wax, and then sent to the Army Medical Museum in Washington, D.C.—which later became a division of the AFIP. There they lay for almost eighty years, until Taubenberger and his team discovered them in 1994.

The next goal was to reconstruct the genes contained in the bits of virus in Roscoe's single lung. But this reconstruction would require millions of copies of the virus, a number far higher than those found in the sample. So Taubenberger had to make copies of the few bits of the viral genes that were left, in the same way that you would photocopy a single sheet of paper. His lab was able to amplify the chains of some fragments of the genes that they found. One of these fragments was the gene that coded for HA, the influenza hemagglutinin that we first discussed in chapter 2. HA, remember, is a critical part of the weaponry of the influenza virus because it allows the particle to recognize the victim's cell, like a radar acquiring its target. HA is more than just a radar, though. Once the virus particle locates and attaches to its target cell, HA then breaches the cell's membrane, like an invading army storming a castle.

Taubenberger set to work with his genetic copier on pieces of the virus that were viable. When he had enough material to analyze, he identified the genetic code that built the hemagglutinin protein on the surface of the 1918 influenza virus and compared it with the genes of other influenza viruses. This piece of genetic detective work—routine today but groundbreaking when it was first performed more than twenty years ago—settled a long debate as to the provenance of 1918. The virus appeared most closely related to a kind of swine flu, although later work would show this strain also had some features in common with bird flu. The official name of the strain would be Influenza A/South Carolina/1/18 (H1N1), after the state from which the sample had originated.

Today, it would take about two weeks to sequence the entire genetic code of 1918, but in the 1990s it took five years for Taubenberger and his team of laboratory workers to identify the complete genome. Along the way, understanding the influenza virus became Taubenberger's professional calling—an accidental by-product of an effort to save his lab from Newt Gingrich's Congress. "It was a gimmick," he said. "I had never taken a class in virology in my life."

Taubenberger's original research identified four fragments of the HA gene segment. Like all genes, they were built out of only four nucleotides, referred to by the letters A, G, C, and T. The building block of 1918, found in one of those fragments, opens like this:

AGTACTCGAAAAGAATGTGACCGTGACACAC

It was the sequence of these four letters, repeated thousands of times across only eight different genes, that turned 1918 into a killing machine. The 1918 influenza virus was composed of various parts, each with a particular role. Some enabled the virus to enter the lung cells; others directed the hijacked cells to reproduce the virus and then to release it, allowing it to infect more victims. When joined together, the virus became deadly.

Taubenberger, together with his lab team, had successfully discovered and then decoded the genetic makeup of the 1918 influenza virus. Later work would use new and faster techniques to verify their findings. But their efforts had been limited by the tiny amounts of raw lung material that they had. They needed more samples to confirm their work, but had exhausted their search among the dusty slides at the AFIP. Help would come from a most unexpected source: a Swedish pathologist who, decades earlier, had failed in his own attempt to find the virus.

*   *   *

In 1949, Johan Hultin came to the United States from Sweden as a visiting medical student. In his twenties and fascinated by influenza, he had taken advantage of a program at the medical school of Uppsala University that allowed its students to spend time abroad. Hultin had chosen to travel to the University of Iowa, both because of its reputation and because of the large community of Swedish immigrants who lived in the area. There he intended to study the reaction of the body to influenza.

In January 1950 Hultin had a chance meeting with Roger Hale, a leading virologist who was visiting from the Brookhaven National Laboratory. Hale, who knew that the Swede was interested in flu research, told Hultin that in order to advance the field he needed actual specimens of the 1918 virus. "We just don't know what caused that flu," Hale told him. "Somebody ought to go to the northern part of the world and try to find a victim of the 1918 flu pandemic buried in the permafrost. That victim is likely to have been frozen since 1918, and you could try to recover the virus."

The conversation quickly moved on to other topics, but the remark made an impression on Hultin. He immediately asked his faculty adviser if he could change the topic of his PhD thesis. Now, instead of studying influenza in the lab, he wanted to get out and hunt for the virus itself. He would find a buried, preserved sample and then analyze it. And that could reveal what made the 1918 influenza virus so deadly.

Hultin was uniquely prepared to search the permafrost. He enjoyed traveling, and before joining the University of Iowa, he had worked in Fairbanks, Alaska, for a German paleontologist named Otto Geist, who gave him free board in return for help digging up mammoth tusks in the Arctic. Now Hultin wanted to return to Alaska, where the bodies of those who had died in the 1918 pandemic were buried in the permafrost. He wrote to Geist and asked if the paleontologist could introduce him to the local Inuit villages and missionaries who worked there. Perhaps Geist could ask those missionaries if they still had records of victims from the 1918 epidemic and where they had been buried. Hultin was only interested in bodies that had been buried in the permafrost, their lungs preserved with intact viruses frozen inside. He also applied for a $10,000 grant from his adoptive university to go find them. That's about $100,000 in today's money—a lot to invest in a visiting foreign student—but the university bought into the harebrained scheme.

Hultin met Geist in Alaska in 1951, and together they traveled to Fairbanks, and then another five hundred miles west to Nome, on the shore of the Bering Sea. Once there they discovered that a local river had changed its course, and during floods had melted the permafrost. There were no soft tissues left, no lungs had been preserved, and so there would be no virus to sample.

Hultin was undeterred. He hired a pilot to fly him to another site, this time farther north: the village of Wales, where almost half of the 400 residents had died in the 1918 epidemic. There was a mass grave, marked by a large cross, that contained the remains of the flu victims. But once again, he found that the permafrost had not been so perma. Hultin persuaded the pilot to fly him on to Brevig Mission, a tiny village on the shore of the Bering Sea, where 72 out of 80 residents died of 1918. But Brevig did not have a landing strip, so he touched down on the beach at a neighboring village, crossed the open water on a whaleboat, and then walked six miles through the soggy tundra.

His tenacious efforts paid off. In Brevig the permafrost was deep enough to have preserved the bodies buried there. In addition, Hultin found three survivors of the 1918 pandemic, whose support would be invaluable. He asked them to describe to the rest of the villagers what it had been like to live through the epidemic, and to witness the deaths of almost everyone during one horrendous week in November 1918. Hultin then explained to the villagers that obtaining a specimen of the virus could create a vaccine and prevent another outbreak. With their support, and that of the village council, he was given permission to proceed.

At first Hultin dug alone. He hacked through the topsoil with a pickax until he hit the permafrost. Using driftwood collected from the beach, he then started a small fire that melted the frozen layer. By the end of his second day he had reached a depth of four feet. There he uncovered the body of a girl, aged about twelve years. The body had been well preserved, and this encouraged him to dig even deeper to find better specimens. He was soon joined by his faculty adviser, a pathologist, and Otto Geist, the paleontologist. At a depth of six feet they found three other bodies. A postmortem exam of the lungs showed they were also preserved, and therefore likely to contain the 1918 virus.

In all they exhumed five bodies, dissected them, and took small cubed samples from their preserved lung tissues. In what seems today to be an insanely reckless move, they wore only gloves and surgical masks for protection. The specimens were transported back to Iowa on dry ice, where Hultin injected mixtures of them into the amniotic fluid of growing chick embryos, the ideal medium for the flu virus. Disappointingly, the virus did not reproduce, so Hultin moved on to live animals—mice, guinea pigs, and ferrets—but was unable to infect any of them with the flu.

It seemed the virus that had killed so many was no longer viable. It had been destroyed by time and the extremes of nature. Eventually Hultin ran out of tissue specimens, and there was nothing more to be done. The expedition ended as a scientific failure. Hultin never finished his PhD. Decades later, though, he'd get a shot at redemption.

For the next forty-six years, Johan Hultin's expedition remained forgotten. He became a pathologist, settled into his career, and continued to travel with his wife. He rebuilt an ancient stone maze in Iceland and hiked in England and Turkey. "I'm going to settle down when I get old," he told a reporter. "I have to do these things now. I'm afraid the warranty will run out."

Hultin, always on the lookout for an adventure, found a new mission in 1997. It was a quest that completed the search for the influenza virus he had started nearly fifty years earlier. While retired and living in San Francisco, he read about Taubenberger's work uncovering some of the genetic sequencing of 1918 from the dusty archives of the AFIP. Intrigued, Hultin wrote to Taubenberger and told him of his 1951 expedition and its disappointing outcome. He offered to fly back to Brevig and try to recover the flu virus a second time. Hultin would finance the mission on his own. It would be a one-man expedition to finish what he'd started.

Hultin had some competition. Around the same time, a thirty-two-year-old geographer at the University of Toronto named Kirsty Duncan was planning her own much larger and well-funded expedition. Duncan was initially interested in the relationship between the flu and the climate, but she also wanted to get her hands on a sample of the 1918 virus itself, to better understand what made it so deadly. She apparently hit upon the idea of an Alaskan expedition independently of Hultin, but was unable to narrow down her search to any known victims. Duncan turned her attention instead to the archipelago of Svalbard, which lies in the frigid sea between Norway and Greenland. Duncan discovered that seven miners had died from the flu in October 1918, soon after they had arrived for work at an outpost called Longyearbyen. If the permafrost had actually done its work, their bodies—along with the 1918 flu virus—would be preserved.

Duncan then put together an international team: the chief of the influenza branch at the CDC, a pediatrician and a geologist from Canada, an American virologist, and Dr. John Oxford from London. Oxford was a virologist with a long-held interest in the influenza virus and, you may recall, had a theory that the 1918 outbreak had originated in northern France.

Duncan was in the midst of preparations when Jeff Taubenberger and his colleagues published their paper that detailed the findings of Private Roscoe's flu virus. Duncan and Taubenberger, unaware of each other's work, met at a workshop in Atlanta that was convened to discuss Taubenberger's reconstruction of the genetic code of the 1918 virus. Taubenberger offered to analyze any samples that Duncan obtained from the bodies of the miners in Svalbard.

But after the publication of Taubenberger's paper, was Duncan's plan still necessary? On the one hand, there was the cost, and the risk that exhuming bodies could expose the expedition—and the rest of the world— to infection. On the other hand, there were scientific concerns that the specimen of Roscoe Vaughan's lung had somehow been altered by its long bath in formaldehyde. If that had happened, it would be critical to obtain additional samples of the virus and compare them to those in Taubenberger's possession. Scientists at the CDC, which had once committed funds to support the expedition, now questioned its purpose. Citing a tight fiscal year, they withdrew from the project, taking their funding with them. Duncan's team still had the support of the National Institutes of Health, and had secured a grant from the pharmaceutical giant Roche. They made the decision to proceed to Svalbard, buoyed by a federal grant for $150,000. There, they would use ground-penetrating radar to locate the bodies, and build a secure biohazard tent over the graves to minimize any risk.

Meanwhile, Johan Hultin, now seventy-two years old, returned to Alaska to dig once again. The village elders in Brevig Mission not only gave him permission to exhume the bodies but also provided four young men to assist him. They dug with pickaxes and shovels and eventually reached a depth of seven feet. And so it was that in August 1997, after three days of digging by hand, Hultin and the four villagers found the body of an obese woman whom he would name Lucy, out of respect for her and the contribution she might make to science. Hultin was photographed kneeling next to a small pile of Lucy's remains, wearing waders and a pair of surgical gloves. Her body fat had insulated her organs when the permafrost occasionally thawed. As a result, her lungs were intact. Hultin removed them, hoping they would contain the 1918 virus, and mailed specimens to Taubenberger using three different carriers to minimize the risk that they would be lost. Within a week, the lab confirmed the presence of 1918 flu particles in them. With more lung tissue than ever to work with, Taubenberger's lab could now rebuild the entire genetic code of the 1918 virus. The *entire* code.

Taubenberger announced the success of Johan Hultin's second expedition in August 1998, just as Kirsty Duncan and her team left for Svalbard and the town of Longyearbyen. Kneeling on the ground and armed with shoe-box-sized ground-penetrating radar, they identified the likely area in which victims had been buried. They dug for eight days inside the biohazard tent before striking the lid of a coffin. The team's elation was tempered by the coffin's location in the upper layer of the permafrost, which meant that the body inside had likely thawed at some point. Out of respect for the dead, the team never publicly discussed the condition of the bodies, though the *New York Times* reported that they had been buried without clothes and were wrapped only in newspapers. Several samples of soft tissue were collected, but none was in a condition to provide viral particles. Duncan returned from Svalbard empty-handed, though she found renown later. In 2015 she became Canada's minister of science in the cabinet of Prime Minister Justin Trudeau.

And the town of Longyearbyen also found later fame. In 2008 it was chosen to house the Global Seed Vault, where seeds from across the world are sent for safekeeping, in case of a global agricultural disaster. The vault is buried five hundred feet below the permafrost, can withstand bomb blasts, and contains more than a quarter of a million species of seeds. Once the site of so much death, Longyearbyen is now a monument to the living, to endurance, and to survival.

And so, thanks to Johan Hultin's intervention, Taubenberger's lab became the sole custodian of samples of the 1918 flu virus. But because each sample might yield only part of the genetic code of the 1918 influenza virus, still more were needed. The search widened, and researchers who had read Taubenberger's original data hunted through their own specimens and slide collections. The Royal London Hospital, which had been founded in 1740, was certainly old enough to have treated patients in the 1918 epidemic. A search through its autopsy archives found two preserved samples of lung tissue as well as the clinical records of the patients to whom they belonged. These records put the preserved specimens into a clinical context that had been missing until now. They described when the patients became ill, how their illnesses progressed, and what the victims looked like as they succumbed to the virus. The records also ensured that the tissue specimens that had been found were taken from victims of the influenza virus itself, and not from a patient who had died due to a secondary bacterial infection.

When Taubenberger compared the genetic fingerprints across all the samples, he found something remarkable. Although they were separated by 7,500 miles (the distance from Brevig Mission to London) and by several months (the earliest sample came from September 1918 and the latest from February 1919), the genetic material of these viruses was 99 percent identical. This suggests that only a single strain of influenza was in circulation in the early stages of the 1918 outbreak, and that just one specific antiviral drug or vaccine might be effective during the most lethal wave of any future flu pandemic.

Today, Jeff Taubenberger continues to hunt for specimens of the 1918 virus that may be preserved in pathology collections around the world. So far he has been unsuccessful, but he retains his trademark optimism. After all, additional samples could further address the question of whether more than one flu strain circulated in 1918 and shed light on how the lethal virus evolved. But sequencing its genetic code cannot by itself help us understand why the 1918 virus was so lethal. It does not tell us *how* the virus worked when it infected or *why* it spread so quickly. To answer these questions, scientists would need to build a brand-new, fully functional copy of the extinct virus.

The resurrection of the 1918 virus took several years of collaboration between the Centers for Disease Control and Prevention, the Mount Sinai School of Medicine in New York, the Armed Forces Institute of Pathology in Maryland, and scientists from the U.S. Department of Agriculture. The actual building of the virus took place in Atlanta, in a CDC biosafety lab where scientists wore breathing hoods. Although the influenza virus was—and continues to be—easily spread between people, to cause illness it must be inhaled. A breathing hood was sufficient protection. In addition, it was thought that the scientists would have a degree of immunity to the 1918 virus, since its flu descendants had been circulating each fall and winter in the interim. At least, that is what they hoped. Just to be sure, those directly working with the virus took prophylactic doses of antiviral medications.

In 2005 the team announced that it had built several versions of the 1918 virus. The first was a fully working clone, with all eight of the original genes of the 1918 flu virus. It was capable of infecting test animals (and humans). The team also rebuilt versions of the virus that contained only one or three or five of the original eight genes, to use as controls. To test how lethal the virus was in mammals, the 1918 virus was sprayed into the noses of mice. Many died within three days. The lungs of these mice contained almost forty thousand times the amount of virus as the lungs of mice infected with the control version. If that wasn't scary enough, the working eight-gene clone turned out to be at least one hundred times deadlier than the five-gene version. With a little more detective work, it became clear that the cause of this turbo-charged virulence was the gene that coded for hemagglutinin (HA), the critical protein that sits on the surface of the virus and attaches it to our cells.

Scientists now had at least a partial explanation of why the 1918 virus was so deadly, but there was still more to learn. One of the clinical features of the 1918 flu pandemic was a bloody, frothy cough that rapidly developed in the victims as the lining of the lungs was eaten away. From looking at specimens of the lungs from the infected mice, it was apparent that the resurrected 1918 virus was able to attract special white cells called neutrophils. These cells are recruited as part of the immune response to fight the virus, but when they go to war they produce a great amount of collateral damage to the healthy lung tissue itself, allowing a secondary bacterial pneumonia to develop. It had long been thought that some of the deaths in 1918 resulted from a "cytokine storm," an overabundance of proteins that play an important role in our immune system. Now for the first time there was evidence that supported this theory.

There was another secret that the 1918 virus gave up in its resurrected state. One of the proteins that the virus manufactured was nearly identical to a protein made by bird flu viruses. This suggested that the 1918 virus did not arise as a result of reassortment, in which a few of its genes traded places with the genes of a bird flu strain. Instead the 1918 virus appeared to be a bird virus that somehow adapted to humans. It also seems to have spent some time in a mammalian host, although we still do not know which one. The 1918 virus traded a few genes with this mammal until it evolved into a perfect killer virus. It had just enough new proteins on its surface to be unrecognizable by our immune system. One of those proteins, HA from a bird virus, caused the body to mount an uncontrolled inflammatory response, destroying its own lung tissues in the process. And the virus generally didn't kill its victims until several days after it infected the lungs, giving it time to replicate in its new victim and to be coughed out into the lungs of others.

When the resurrection was revealed in the October 2005 issue of *Science*, scientists were shocked and alarmed. Was the published paper too detailed in its description of the process? It is standard practice for scientists to share their experiments and results; this allows others to replicate and verify the original experiments, and it buoys the authors' reputations. But wouldn't information on how to resurrect the deadly virus be dangerous if it fell into the wrong hands?

The newly active 1918 virus rekindled a debate over "dual use" information. These new flu details could be used to create vaccines and treatments, prevent a repeat pandemic, and improve the health of our civilization. Or they could be used for nefarious purposes: hostile governments or terrorist groups could perhaps weaponize the flu. A great deal of scientific information is dual use, meaning it can be used for both good and evil. When physicists first split the atom in 1939, they realized nuclear energy could be used to either power a city (with a generating plant) or destroy it (with a bomb).

Before the re-creation of the pandemic influenza virus, there had been another dual-use controversy that began in 2002. Scientists from Stony Brook University announced that they had produced a polio virus from scratch by using a map of the virus that was available online and purchasing the chemical building blocks through a mail-order company. But could this information also allow a fanatic to build copies of polio without having access to a natural virus? What if terrorists used this methodology to build a highly contagious virus, like Ebola?

An attempt to resolve the dual-use question began in 2005, when the National Academies of Science appointed a committee to address it. After much deliberation and a report called *Biotechnology Research in an Age of Terrorism*, the committee recommended that the scientific community police itself. In an age of global information sharing, it made little sense to regulate papers published in the United States alone, because the authors would simply turn to a journal in another country with less stringent rules. To help scientists with their task of self-regulation, the committee also recommended the appointment of a National Science Advisory Board for Biosecurity (NSABB), which could provide advice and guidance.

Donald Kennedy, the editor of the journal that published the resurrection paper in 2005, had to grapple with the implications of having done so. Would the instructions on how to build the virus fall into the wrong hands and lead to a mass dispersal in a crowd at a football game, in a mall, or in the subway? Would the resurrection of 1918 lead to a *repeat* of 1918?

Before publishing, he sought advice from officials at the CDC and the NIH, all of whom supported publication. At the eleventh hour, Michael Leavitt, the secretary for the U.S. Department of Health and Human Services, insisted on getting approval from the NSABB. The paper went to press without the approval. Kennedy stuck by his decision, noting that the government "can't order the nonpublication of a paper just because they consider the findings 'sensitive.'"

Scientists are by definition a curious bunch, and now they had a sample of the 1918 virus to tinker with. What would happen if they added one kind of influenza gene or removed another? Would the virus become more or less deadly? Over the next several years the scientific community continued to study not only the 1918 virus but also several other pandemic flu viruses. For example, the H5N1 virus does not naturally have the ability to spread via droplets in humans, but it certainly could evolve that capability by the natural process of gene reassortment that happens in the wild. What would happen to the virus then? Would it become more deadly, as might be expected? Or would there be an unexpected genetic interaction within the virus that rendered it less dangerous?

There was only one way to find out. In 2012 an international group genetically modified the H5N1 virus and infected ferrets with it. It soon mutated and became airborne, but to everyone's surprise, it also became *less* deadly. In another experiment, researchers at the University of Wisconsin took an avian flu virus that was similar to the 1918 influenza virus and tinkered with its genes just a little. When this virus was tested on mice, it proved to be more lethal than the original avian virus from which it originated.

All this tinkering was creating superviruses that did not exist outside the lab and that might be more easily transmissible between different species, or more virulent, or more resistant to any influenza vaccine. Most researchers were insistent that these "gain of function" studies were needed to better understand how the flu virus might evolve, but the federal government saw things differently. These experiments were a security risk.

In October 2014 the White House paused the federal funding of gain-of-function experiments to assess the risks and benefits. Many of the genetic experiments on the 1918 flu virus and its descendants came to a halt as the scientific community debated the wisdom of proceeding. Vaccine researcher Peter Hale thought the pause was an excellent idea. "The government has finally seen the light," he said. "This is what we have all been waiting for and campaigning for. I shall sleep better tonight."

Others thought that the pause was unnecessary and that it would stymie important research. It continued through January 2017, when the White House released new research guidelines. Any experiment that involved the creation of new viruses would need to be reviewed by an outside panel of experts and defended by the researcher. But these guidelines were not implemented, and so the ban on gain-of-function research remained in place. Then, in a move that surprised many, the government lifted the ban in December 2017. It released a brand-new set of rules to guide funding decisions that involved research on influenza, SARS, Ebola, and other dangerous viruses. These rules also covered research on viruses with a gain of function, and with their release the NIH promptly removed its ban on funding this kind of research.

Thanks to Johan Hultin and Jeff Taubenberger, we now know intimate details about the 1918 virus, including the sequences of its genetic building blocks. However, Taubenberger believes we still have far to go. He points out that we still don't know why flu strains can affect some mammals but not others. We still don't know if 1918 was a reassortment of an existing flu strain that turned suddenly lethal, or if it was a novel virus that appeared as if from nowhere. We still don't know why 1918 was particularly lethal in young adults, the very group that is usually most resilient to these kinds of infections. We still don't know what happened to the flu virus in the years after the 1918 pandemic—where it went and why it became less lethal. So many unknowns despite so much new information.

"I've been thinking heavily about the flu for twenty years," said Taubenberger, the man who knows more about the flu than anyone, "and I know nothing."

6

DATA, INTUITION,
AND OTHER WEAPONS OF WAR

In the emergency room, we aren't focused on the many things we don't know about the flu virus. We are simply too busy handling the cases of flu lying on the gurneys in front of us. The questions that an ER doctor focuses on are these: Does the patient coughing, aching, and sneezing in front of me have influenza? Do I need to treat with medications? Do I need to admit the patient to the hospital?

Most ER doctors, including myself, don't usually bother to get a rapid flu test for a patient but rely instead on the patient's story and symptoms. If a patient has chills and a runny nose, if she is tired and has a fever and night sweats, if she aches as if hit by a truck, and if it is late fall and her roommate had the same symptoms just a week ago, then she most likely has a viral illness that is influenza, or something very similar. And that's a good enough diagnosis, because most doctors remember what they were taught in medical school: if the outcome of a laboratory test will make no difference as to how you will treat the patient, don't bother to order it.

Nearly all the influenza patients I see in the emergency room are well enough to be sent home, with advice to take some over-the-counter medicines to control the fever and body aches, rest, and drink plenty of fluids. This treatment plan will not change in the slightest if I have a laboratory test confirming my clinical diagnosis. And if it isn't actually the influenza virus at work but one of a dozen others that produce an influenza-like illness (ILI), the result will be the same. The patient is still going to be discharged, is still going to take Tylenol or Motrin for the fevers and body aches, and is still going to rest at home and drink plenty of fluids. Since this is the case, I almost never order a rapid influenza test; its result will be of no consequence to the way I will treat the patient. For these reasons, *not* getting a flu test is often good clinical practice.

Even if your doctor gets a flu swab, that tells only her about you. It doesn't tell the local health department about the number of influenza cases that are out there. And this is important information because it allows for community-wide planning and, if needed, for special measures to be taken, like closing schools. For that to happen, the data from each patient needs to be reported. This is its own major challenge. Data collection relies on the goodwill of hospitals and their staff, who—in addition to their already heavy workload—must review their day's work and fill out forms about the number of flu cases they've treated. Will the reporting always happen, and will it happen both promptly and completely? There will be double counting; if a patient with influenza is seen in the emergency room and then admitted to the hospital, should she be counted among the ER statistics or the inpatient statistics (or both)? Also, who should fill out the forms? Nurses? Doctors? Physician assistants? As with all tests, someone needs to pay for the swab and the laboratory materials and for the technician to put the results into the computer. Performing this test for surveillance purposes on the tens of thousands of people across the country with a flu-like illness can cost many millions of dollars.

Many states request—but do not require—that their primary care doctors, pediatricians, internists, and urgent care clinics track the number of patients with flu-like symptoms. In California about 150 providers do this. But in sunny Florida only 43 providers take part in their state's influenza tracking system. That's 43 providers for a state of almost 20 million people. Since the CDC recommends one provider submit data for every 250,000 people, Florida has half the number needed to get useful statistics. Relying on the goodwill and volunteer spirit of health care providers—people who likely already have difficulty keeping up with the demands of patient care—means that the flu data they deliver may not always be timely or even complete. We're left with a patchwork of information that is, in a sense, useless for big-picture planning.

In an average eight-hour ER shift I would see between thirty and fifty new patients, and in the fall and winter perhaps ten or more would have symptoms of the flu: cough, fevers, body aches and chills, fatigue and sweats, a runny nose, and a sore throat. Some patients might have only one of these, while others may also come in with vomiting—or perhaps *only* with vomiting. Now I am asked to estimate how many of these patients had influenza. Do I report all the patients with aches *or* chills *or* a fever, or only the ones who had aches *and* chills *and* a fever? What if they had a fever and vomiting, but no body aches? Without a lab test to diagnose the precise virus that is making my patient miserable, I cannot be sure whether I'm seeing true influenza or one of the many ILIs that mimic its symptoms. My clinical judgment can take me only as far as a rather imprecise final diagnosis: "fever" or "viral syndrome." If I diagnose viral influenza, I may be right only part of the time. Which is fine, in terms of patient care. But it's not fine if you're trying to collect data on the severity of the influenza season.

There are a lot of viruses out there to make you sick. There is the rhinovirus, which causes the common cold, and the family of rotaviruses, which cause nausea, vomiting, and diarrhea. There is the adenovirus, which causes conjunctivitis, coughs, a runny nose, and body aches. The human respiratory syncytial virus usually infects young children, giving them a fever, a nasty cough, and a runny nose. But none of these is the influenza virus, which is the only one we want to track. Without a test, your primary care physician cannot tell you which is causing your symptoms. And as we've seen, it doesn't matter to the patient. The treatment is the same for all, so there is no need to get a costly lab test. However, if you are an epidemiologist who wants to predict where and when the next influenza epidemic might erupt, you must track influenza proper. You cannot rely on these clinical diagnoses of flu. Using lab tests to distinguish between ILIs and true cases of influenza becomes unavoidable. In the emergency room I would almost never obtain a flu swab, for all the reasons we've just outlined, but every now and again our faculty wanted to know if "the flu" was really *influenza*, and for a while I'd find myself swabbing patients—using a small tool in an attempt to answer a big data question.

Except sometimes data collection can backfire. Over the summer of 1992 in Fairbanks, Alaska, the public health laboratory received nine positive flu swabs from the office of a single primary care doctor. They were all from children younger than nine years old, but since children are often the first to become ill during the flu season, that was not unusual. What was unusual was that the cases had been detected in the summer, outside of the usual flu season. This uptick in cases came to the attention of the CDC, which sent an agent to Fairbanks in an attempt to determine if the cases might represent the beginning of a fresh flu outbreak. The agent was Ali Khan, now dean of the College of Public Health at the University of Nebraska Medical Center, but back then a medical epidemiologist working for the CDC. Khan was concerned that the influenza might be a pandemic strain. After all, the 1918 pandemic came in two waves, the first of which appeared unseasonably, over the spring and summer.

The reason that you haven't heard of the Fairbanks influenza epidemic of 1992 is because there wasn't one. Khan was dispatched to

Alaska because one pediatrician was a little too meticulous and sent a flu swab from any patient with a runny nose. Influenza naturally circulates at low levels in the summer, and the pediatrician's thorough testing merely reflected the usual amount of flu; it was just being reported from a single office. Overall, there wasn't any more influenza around than usual. The false alarm was caused by data.

We live in an era when the answers to so many questions are just a Google search away. Where should I go for dinner? How much is a flight to Santa Fe? What's the French translation of "Do you have cold medication?"

Do I have the flu?

Go ahead. Google that. More than 1.5 million results will pop into your browser in less than a second. You may see a hit—sponsored by Tylenol—that says "Feeling under the weather? Flu activity is high in your area." You may see a link to WebMD: "Flu or cold? Know the difference." In an earlier era—say the winter of 2008—Google would have scooped up your flu-related queries. Your question would have become Google's answer to when and where the flu was spreading.

Google's foray into flu prediction began in 2008 with a new service: Google Flu Trends. First, Google went back and looked at billions of searches that had been performed over the previous five years. Every year in the United States at least 90 million adults search Google for medical information, and Google hunted for flu-related queries ("cough," say, or "chills") and matched them with the CDC's historical flu data. Google then applied those queries to predict what might happen going forward in time. On January 28, 2008, for example, flu queries spiked on Google Flu Trends. Two weeks later, the CDC itself reported an increase in infections.

Silicon Valley had produced real-time flu data that was beyond the reach of slower-moving hospitals, scientists, and the medical bureaucracy. If its algorithms were accurate, Google Flu Trends could help the government and medical industry prepare properly and respond early during a flu season.

A major task in trying to prevent or curb a flu outbreak is figuring out who exactly has the flu. As we have seen, this is much trickier than it sounds. And Google Flu Trends appeared to be a solution, or at least a powerful tool using big data, which seemed to provide a depth, breadth, and complexity of information that nurses and doctors were incapable of collecting, let alone processing. Computers had leveraged a simple Google search to produce something quite sophisticated: an estimate of the influenza burden. Or so it seemed.

For a while, it looked like Google Flu Trends was a smash hit. It seemed to predict flu in Canada, Australia, and several European countries, and was corroborated with data obtained from the sale of antiviral medications. However, there was a hiccup in 2009, when Google Flu Trends underestimated an outbreak of influenza A in the United States. The algorithm was updated to include more search terms directly related to the flu and fewer that had to do with its complications. That hiccup was a harbinger of things to come. The winter of 2012 proved fatal for the program.

That year the U.S. flu season involved a rather virulent strain, and people became sick and died at higher rates than were typically seen. But when it was all over, it turned out that Google had overestimated the already high amount of influenza by as much as 50 percent.

What went wrong? Perhaps the algorithm was clunky. It had to be calibrated anew each year, and was therefore never as refined as it should have been. Or perhaps the problem was more fundamental, and had to do with Google itself. The company's core mission, after all, is not to provide the best data it can on the prevalence of flu. Instead, Google is a corporation with profit as its highest priority, and much of its central business model is to generate advertising revenue from its powerful search engine. In the influential journal *Science*, it was suggested that some of the tweaks that Google made to the algorithm were done to improve its business model, and that this came at the expense of its predictive accuracy.

Perhaps many people searched for key flu terms on Google not because they were ill but because they were *afraid of getting* ill. These "worried well" internet users never became sick, but their searches remained part of Google's data set. Remember that the 2012 flu was particularly nasty. The media reported its severity and New York declared a public health emergency. Perhaps this drove up the number of people Googling "the flu," which of course did not equal the number of people *getting* the flu. In the end, Google Flu Trends was quite accurate in at least one respect: quantifying the peaks and valleys of its users' *interest* in the flu.

Underneath all this was perhaps a bit of hubris. Big data does indeed allow an unprecedented look at millions of data points, but those points don't always reflect an accurate picture of the ground level. Recall the 2016 presidential election, in which nearly every data point signaled a victory for Hillary Clinton. Or consider Boston's Street Bump app, which uses the accelerometer in your smartphone to detect potholes in the road. The city would learn where the potholes were from crowdsourcing; a citizens' army would automatically send data on their location. But the app only produced a map of where young, affluent car owners—those who would typically download an app like Street Bump—were driving over potholes. And while that specific set of data was very complete, it did not reflect the location of *all* the potholes in Boston, just as political polling did not reflect what was going to happen to Hillary Clinton in Michigan, Wisconsin, and Pennsylvania. The circumstances were broader and more complex on the ground.

Did this mean that vanguard technology was worthless without the steadying force of more traditional and time-tested methods? There we were, ninety-five years after the great flu epidemic, and yet the dazzling evolution in technology was still not enough for us to get a firm grasp on who had the virus. Perhaps a mix of traditional and innovative tools— throat swabs and algorithms—would give us our best shot at identifying the infected and curbing epidemics.

"It is hard to think today that one can provide disease surveillance without existing systems," said Alain-Jacques Valleron, the French epidemiologist who in 1984 founded the first computerized flu tracking program. "The new systems depend too much on old existing ones to be able to live without them."

Some predicted that Google would once again update the program and refine its algorithm, but in August 2015 its flu team wrote a goodbye note of sorts. They discontinued their own website and instead began to "empower institutions"—like schools of public health and the CDC itself—to use data to construct their own models.

We need to monitor influenza numbers for several reasons. Without an accurate count, it becomes impossible to track the advance and retreat of influenza. Health departments need accurate numbers to prepare, whether by stocking up on vaccines or advising the public about the flu risk. Vaccine producers look back at the influenza count to determine if that year's vaccine had been the correct one. And knowing the precise strain that is circulating one year is crucial to predicting what strain will be in circulation the next. So if the power of big data and the tentacles of technology can't count the cases of influenza, what can?

Around the same time that Google Flu Trends was getting started, a group of economists led by Forrest Nelson at the University of Iowa tried a different approach to estimate the burden of influenza. Nelson had spent years on the boundary of where prediction and economics overlap: the stock market. When we buy a company's stock, we believe the company will grow, outperform the competition, and produce profits. The more people believe in the future success of the company, the higher its stock price will be. Conversely, if we judge that a company's economic future is bleak, that it will not succeed, its stock price will fall as shareholders scramble to sell.

Nelson applied the prediction market to politics, forecasting the outcomes of elections, and then turned to influenza. Could he harness all kinds of disparate expert knowledge to predict how much flu there would be? That question gave birth to the Iowa Flu Prediction Market. Nelson wanted to draw from a wide range of flu-savvy people in the state: nurses, school principals, pharmacists, physicians, and microbiologists. Their aggregate information would, he hoped, provide insight into how much flu there was, and allow him to predict how much there might be in the future.

Nelson started modestly in January 2004 by inviting fifty-two health care workers with a variety of backgrounds to play. He had secured some grant funding and gave each of his traders fifty dollars to use. With this they bought and sold contracts based on a map the CDC would later release to visualize influenza activity. For example, in January you could buy a contract for the first week of February that called for widespread activity in Iowa, and that would be shown in red on the CDC's influenza map. Or perhaps you believed, based on all the information you had at hand, that the influenza activity was only going to be sporadic (shown in green on the map). Then you'd buy a contract for that color. A contract that denoted the actual level of influenza when it was finally released by the CDC was worth one dollar. All the others were worthless.

The Iowa Flu Prediction Market continued over several flu seasons, and had some early success. It correctly forecast the CDC's official influenza count as often as 90 percent of the time, though that number fell the further out the predictions were made. But there were also problems. It was challenging to get enough physicians interested; most told Nelson they simply did not have the time to trade. He ran out of funding and could no longer provide real cash. So he switched to fake money, in the form of "flu dollars," and continued the project. But those playing the market seemed to tire of using pretend money, and participation dropped off. One of Nelson's co-researchers passed away, another moved on to other research areas, and in 2012 the Iowa Flu Prediction Market ceased trading.

When I spoke with Nelson, he had retired and was enjoying the warmer climes of Austin, Texas. "It was a baby of mine since 1988," he told me, referring to the year his presidential prediction markets had first opened. He acknowledged that running the flu market was costly in terms of both time and money, and he was frustrated that there hadn't been a greater buy-in from the medical profession. But he had never expected the prediction market to replace traditional influenza surveillance. Instead it could be used as a supplement, offering another data point for public health officials to use. And, like any parent, he remained very proud of his baby.

Google searches and doctors' reports all flow toward one agency inside the CDC: the National Center for Immunization and Respiratory Diseases, in Atlanta. Within that center sits the influenza division, whose staff of three hundred must predict, track, and recommend the treatment for influenza using the data at hand—some of it helpful, some of it flawed, and some a mix of both.

The division relies on the work of clinical laboratories, like the one at my hospital in Washington, D.C., and of public health laboratories, like the one in Fairbanks. Every week across the United States about two thousand health providers—nurses, physicians, and their assistants—fill out a form that tells the CDC how many patients they've seen with an influenza-like illness. It's a time-consuming but valuable report from the front line of the battle against influenza, but it has obvious limitations in terms of the quality of data it produces. Remember that one doctor might report "influenza" while another doctor seeing similar symptoms might report "fever," or "gastroenteritis" or "viral syndrome," all ILIs. When it comes time to aggregate the numbers and report ILI activity to the CDC, the electronic medical record might include some, all, or none of these diagnoses.

The CDC also relies on hospital labs to report how many tests for influenza they run, and how many of these are positive. You might think that these data would be more accurate than reviewing the electronic medical record, but here, too, the true incidence of influenza might vary, depending on which patients were swabbed and on the locations of the clinics and hospitals. You might have a doctor who swabs only when she is treating a patient who is very ill, or one who has cancer, HIV, or another complicating condition. In this case, the pool of all patients swabbed is limited but the number of positive cases is high. Or you could have the inverse: another doctor—even in the same hospital—whose practice is to swab many patients, not just those with a chronic illness. In this case, the sample size would be quite large and the number of positive cases of influenza comparatively small. And these numbers, in either situation, include only those who choose to see a doctor, and those doctors who choose to swab their patients. This imperfect, sometimes contradictory information is what the CDC has to work with.

And it's solely *reactionary*. The numbers show only what has already happened. The time lag between the collection of data and reporting it to the public can be days, or weeks, or longer. The data might indicate the *burden* of flu (the impact in a given place) but it lags behind the *prevalence* of flu—that is, how much influenza is actually circulating. It tells us what was, not what is or what will be. For example, if I see three patients with influenza in the first week of November, nine in the second, and thirty in the third week, I could reasonably estimate that in the last week of November I may see as many as seventy new cases. With this knowledge, I would prepare my clinic for an outbreak. But this data may not predict an increase at all. Perhaps the flu epidemic *peaked* in the third week, and the number of new cases will begin to fall. If that is the true scenario, then I prepared my clinic for a rush that never came.

This is the situation we are in as I write these words. The first weeks of January 2018 had a sudden and dramatic increase in the number of confirmed cases of influenza. Had we reached a peak, or would the numbers continue to climb? Nobody knew. The press, meanwhile, continued to cast the data as an influenza pandemic, forgetting the lesson of the 2009 swine flu outbreak. When that earlier outbreak was all over, the actual number of deaths from influenza was lower than during a regular flu season.

Counting and predicting influenza activity is really hard. Google Flu Trends tried and failed, and there were no insights from the now defunct flu prediction markets. Data coming from clinics and labs is incomplete and sometimes misleading. So what else might work?

One approach is to skip the data from hospitals and doctors altogether, and focus more on the population of patients. Only a minority of patients with flu-like symptoms pull themselves out of bed and schlep over to their doctor or local emergency room, so you'd have to find a way to reach the majority who stay home or emerge only to get over-the-counter medication. National pharmacy chains like Rite Aid or CVS have data on how many units of flu medicine they sold yesterday, or in the prior week. This data is available in real time with near-perfect accuracy; it does not rely on the subjective entry of a diagnosis or on the decision to obtain a flu swab, but instead links the register scanning your purchase with a database of products that are bought when the flu is in the air. It does not distinguish between real influenza and ILIs, but the two usually rise and fall in harmony.

In fact, the New York City Department of Health has employed exactly this strategy to rapidly detect outbreaks of influenza. The department's efforts began in 1996, when it focused on surveillance for the nasty waterborne diseases that cause gastroenteritis. The program started by receiving weekly reports of the sales of antidiarrheal medicines, and soon expanded to track medications for ILIs. The department had quite a task, since it estimated there were at least 400 different drugs sold in the "cold department." Fortunately, it was able to narrow this down to the most popular medications, the 50 or so drugs that had the terms "flu" or "tussin" in their descriptions. The program also received data in real time; nearly all pharmacy sales were reported to the health department by the next day.

But when the department reviewed its performance in detecting early influenza over a three-year period, it was disappointed. Although the medication monitoring system mirrored the natural rise and fall of flu cases over the autumn and winter months, it could not detect any *early* signal of flu. Just why is unclear: Perhaps people bought medications early, just in case, before the onset of a flu that never came. Perhaps the same medication was used by multiple members of a family, so that the purchase of one unit did not represent the illness of just one individual. Whatever the reason, once again this approach—an early use of big data to detect flu outbreaks sooner than using conventional methods—didn't deliver. Despite this, the city health department has recently stepped up its pharmacy surveillance program, and now monitors both over-the-counter and prescription medications for colds and flu. They've also expanded to include pharmacies outside of Manhattan. The Department of Health now knows how many residents of Queens and Brooklyn are buying cough syrup or cold remedies.

The state of Maryland had another idea involving the public. In

2008 it enlisted a people's army of influenza trackers. As part of the Maryland Resident Influenza Tracking Survey (MRITS), citizens voluntarily sign up on a website hosted by the state's Department of Health and Mental Hygiene. Once a week they answer a couple of simple questions about whether they or any members of their household have flu-like symptoms. This data comes straight from the source and relies only on the presence of symptoms, so there's no need to analyze how many bottles of flu medicine were sold or how many positive flu swabs were received in a lab. In its first year, more than 500 residents of Maryland signed up to participate, and nearly half responded to a reminder email each week. Since then, the program has grown to over 2,600 participants.

I am one of them. Once a week there is an email waiting for me in my inbox. If no one in my household has a cough, fever, or sore throat, there's a simple link to click on, and I am done. It takes about two seconds. It takes only a little longer to report a family member who has flu-like symptoms. Then MRITS asks me if that person sought medical treatment for their symptoms, traveled in the week before they became ill, or missed their usual daily activities as a result.

Although not everyone remembers to fill out the weekly review, the data produced is pretty close to the data from other surveillance methods. For the 2014–2015 Maryland influenza season, for example, the incidence of ILI symptoms reported voluntarily by medical professionals was 1.6 percent; emergency departments in the state reported an incidence of 2.3 percent; and the citizen-driven MRITS incidence was, like Goldilocks, right in the middle of these: 1.9 percent.

The MRITS network suffers from the same limitations we've already noted. Residents report symptoms that are not caused only by influenza. And it is driven by people who are keen to help—so keen that they somehow found out about the survey, signed up for it online, and reported on their household symptoms weekly. How typical of the general population is this self-selected group of flu watchers? Are they like the users of the Boston Street Bump app? Does this group get flu-like symptoms at a lower or higher rate than do their fellow residents of Maryland?

We don't know the answers to these questions. But we do know that some citizens go way beyond self-reporting symptoms. They are so enthusiastic about the flu that they turn it into a vocation, an avenue for amateur sleuthing and study. Grassroots flu groups are all over the internet. Some are blogs maintained by one person, often with a very specific agenda, and others seem to be more objective, providing flu details without any editorializing. Could their size and agility accomplish certain things that major tech companies and bulky bureaucracies could not?

From her home in Winter Park, Florida, Sharon Sanders is the editor in chief of FluTrackers.com. Her website is rudimentary but sprawling, with dozens of old-school chat forums devoted to influenza and other infectious diseases. She conceived it in 2005, around the time that President George W. Bush was reading a history of the flu during his summer vacation, and the look and design of the site hasn't changed much since. Sanders has no medical background but became fascinated with the flu when, many years ago, she saw a TV segment by Sanjay Gupta, the medical correspondent for CNN. He had just visited the CDC in Atlanta, and explained how influenza epidemics are cyclical. Sanders had not heard of repeating influenza epidemics before, so she did what inquisitive people do to find out more. She Googled it. (If she'd done so a few years later, her queries would've been false alarms scooped up regardless by Google Flu Trends.)

She found two (now defunct) discussion sites, Flu Wiki and CurEvents, on which there were rigorous conversations about every aspect of pandemic influenza: preparations, health care workers, 1918 data, medical considerations, and traditional medicine. Sanders recalled one especially animated discussion thread about whether migratory wild birds could spread flu. Sides were drawn depending on whether you were a wildlife supporter or had a scientific background. It got ugly. After a while the discussions devolved into debates about guns for personal protection and other diversions not related to influenza. Sanders had enough, but by now she was hooked on the subject.

"It became clear that starting a new site would be the only way to have a more serious online environment," she says. "So we did." She had made online friendships with a few people she had met through CurEvents. Together with two fellow enthusiasts—one a software engineer and the other a botanist—she launched FluTrackers in February 2006. "We were just actually regular people, concerned citizens without any medical background," she adds.

Now, I'm all in favor of citizen efforts to keep us informed, but for a project as big as Sanders's, wasn't some kind of quality control advisable? She saw things differently, and believed there was a great benefit to unleashing what she calls "previously untapped talent" among the general population. Sanders loads the site with the latest influenza information from the CDC and the World Health Organization. "Our rules were simple," she says. "No bashing, no violent talk, no politics, no religion debates. Respect for others. We were a boutique site for the small amount of people online who wanted to explore disease spread, particularly the flu. . . . Our discussions were serious. And it was fun."

Within weeks of the site's founding, several scientists joined, most of them anonymously. Sanders could vouch for their credentials from their email addresses. Some posted news and scientific papers they thought others might enjoy. Journalists who covered the flu also joined, though nearly all used pseudonyms. Then, as now, anonymity was really important for those who visited this and many other sites. The defunct Flu Wiki's founders remained anonymous for years. Today, almost all of the professionals and new members who sign into FluTrackers do so anonymously.

Over time the web traffic grew and FluTrackers expanded its focus to include other infectious diseases. Tracking influenza outbreaks is only one of the missions of the site. It also reports on new academic papers, conference notes, and expert speeches. But one of the most useful features is its global reach: the site logged almost 18 million page views in the first ten months of 2017. Eighteen million! Who knew influenza was so popular? FluTrackers not only gathers information but also helps figure out how to use it to inform the general public. Since it has such a large readership, it was even invited to take part in tabletop exercises run by the U.S. Department of Health and Human Services. These exercises evaluated how online media could help disseminate information to the public during an influenza pandemic. For a homegrown clearinghouse on all things flu, this was quite an achievement. Sanders thought so too. "I know it seems improbable that an international group of hobbyist volunteers who have never met in person could rise to such prominence," she admits to me. "But over the years this is what happened."

FluTrackers translates foreign news articles and is used by the CDC, the WHO, and a host of other organizations. In an email, Sanders told me that "an alphabet soup of U.S. government agencies view FluTrackers daily to see what we have found." Because its member base is international and on the ground, the site is often able to report disease outbreaks before they come to the attention of larger, less nimble organizations. Sanders "specializes" in Chinese and Arabic sources, which she deciphers with the help of machine translators. She has also learned to look for indicators of flu activity, which are specific to each country. This is especially needed in countries in which the media is tightly controlled. For example, Sanders once learned that in a particular province in Egypt, health care workers were distributing pamphlets on H5N1 flu in a door-to-door campaign, which likely indicated an uptick of flu cases there. Sanders was also tracking the "amazingly frequent" reports of poultry farms being destroyed by electrical fires in Egypt. Since the government did not compensate farmers for the loss of chickens from bird flu, she suspected—though she had no firm evidence—that some of them were staging these fires to obtain insurance payouts and protect themselves from financial ruin. In one province alone, three poultry farm fires were recorded on the same day. So the more poultry farm fires reported in the Egyptian media, the greater the likelihood of an uptick in bird influenza.

Indicators such as these are of special interest to FluTrackers because they may provide a clue as to when new outbreaks of influenza will occur. Sanders compared this research with the CDC or WHO influenza reports, which are indicators of where flu was, not where it might be going. Of course, there is the possibility of error, but that is something she has come to accept. "We could be wrong," she says, "but we are wrong in a very earnest way."

While FluTrackers has no fact-checkers, anyone who posts is required to provide a link to the original news source, unless it would be dangerous for them to do so. This danger is a reality for some who upload information to the web in countries like Egypt and China, where the news media is tightly controlled.

Sanders told me that unlike many blog sites that focus on infectious diseases, her site is apolitical, has no agenda, and is just in the business of getting information out to the public. And the cost of getting this information out is a mere fifty dollars each month in internet service fees. She has great pride in the fact that the site takes no money from business, government, or those with an agenda.

Sanders hasn't yet formed an opinion as to whether there will be another 1918-like pandemic soon. She notes that there are more strains of humanized novel flu than ever before, but whether this portends a new flu epidemic is unclear. She is surprised that there have not been pandemics from new strains of bird flu that were found in Southeast Asia. Some of them had a nerve-racking death rate of more than 50 percent. Closer to home, she is critical of the lack of attention paid to pandemic preparation over the last decade, and saddened by how many flu experts have retired from public service. This has left a lack of knowledge that she fears will tremendously weaken any future federal pandemic response.

FluTrackers is impressive, but has limitations as well as oversight issues. The site reports on where influenza has been suspected but not always where it has been confirmed. The challenge for those in public health is to know what to do with the enormous amount of information the site gathers. Do reports of pneumonia indicate an uptick in complications from an influenza infection? If the Egyptian media is reporting an increase in poultry farm fires, how are we to act on this information? Should we make more vaccine against the last known avian flu that infected us? Or should we instead get more data before ramping up vaccine production? There is a limit to what data points can really tell us. Often, they yield only more questions.

Organizations like the CDC or the WHO are still the best destinations for data on the year-over-year rise or fall of the flu. This information, together with the number of vaccinations, also gives us a measure of the success of preventive efforts. Depending on how quickly the counts are made, a state or town might employ that data to help health officials target their public messaging.

In spite of all this, we still don't have an accurate way to measure how many cases of flu there are each season. We can't solely rely on huge data-driven companies like Google to figure it out for us, nor on citizen-led efforts, and even the CDC data is limited. The influenza virus, a most primitive organism, seems to run circles around our advanced technology. We don't even know the answer to one of the most important questions about influenza: Why does it rise and fall with the changing seasons?

7

YOUR EVENING FLU FORECAST

I first heard about the full moon effect while training in Boston as an emergency physician. The theory is that more patients with psychiatric problems come to the emergency room when the moon is full, compared to when it is waxing or waning. The relationship between mental health and the moon made its way into our language long ago. The word *lunatic* is derived from the Latin *lunaticus*, meaning "struck by the moon." This full moon effect has even been the subject of academic study. At least five research teams have investigated the correlation between the phases of the moon and visits to the emergency room. None found any evidence of association, though try telling that to the ER staff on a busy night when a packed ambulance bay is illuminated by the gray cast of the full moon.

But there *is* a rhythm and flow that is fairly predictable in an emergency room. It's usually quiet until the late morning. The busiest times are from about noon to ten p.m. As noted earlier, on Thanksgiving and Christmas Day very few patients come in. Black Friday is just the opposite; the ER overflows with patients. When I worked in inner-city emergency departments, summertime was stabbing season. There was more happening outside, more places where people gathered, more alcohol flowing. "If it is hot enough to barbecue," a patient of mine once told me as I sutured the knife wounds on his chest, "it's hot enough to stab someone."

Influenza is predictable, too, and that's part of its mystery. It hits us with clockwork regularity, but we don't know why. It appears in the fall and winter and then, like a hibernating bear in reverse, disappears in the spring. Other infectious diseases have a seasonality too. Polio outbreaks were seen in the summer months. Measles would also rise and fall with the seasons. These diseases have been virtually eradicated by vaccines, but the flu remains a ritual, which can lull us into a false sense of security. That's what happened at the onset of the 1918 pandemic.

In September 1918 the *Journal of the American Medical Association* noted that a new and more virulent form of influenza had broken out in several American cities and many army camps. Experts recommended calm. Outbreaks appeared to be following the usual pattern of influenza, and the first wave had already practically disappeared from the Allied troops. Since influenza was seasonal, it would be gone by the spring.

"It should not cause any greater importance to be attached to it," the *Journal* said, nor "arouse any greater fear than would influenza without the new name." The pattern of the 1918 pandemic was fairly typical of the flu. But the number of lives it claimed was not.

Why is there a flu season? Why is it unusual to get the flu in the summer? The spread of influenza has long been suspected to hinge on the climate. For example, some scientists have suggested that the way in which air flows on the outer edge of our atmosphere could play a role in the seasonal appearance of the flu, pushing more virus particles down into the air we breathe. Beyond these atmospheric changes, other researchers have focused on the role that humidity plays in the seasonality of the flu.

It should be noted that the seasonality of flu cases does not happen everywhere. In the tropics, there is no such thing as a flu season. There, influenza is generally present at low levels throughout the year, although in some places it spikes during the rainy season. Only in temperate regions do flu outbreaks rise and fall with the seasons. These regions are north and south of the tropics, extending as far as the Arctic and Antarctic Circles. Temperatures vary widely within Europe, Canada, and the United States, as well as much of Russia, North Africa, and the southern tip of Australia. The farther from the tropics, the bigger the change in climate between winter and summer, and the greater the seasonality of the flu virus.

There are several possible explanations for why there is an influenza season. One of the best-known has to do with how we crowd together: it's called the "indoor contagion theory." In the winter, the theory goes, people spend more time indoors, and there is a greater likelihood of close person-to-person contact. Coziness and proximity encourage transmission of the virus, and the number of influenza cases rises. This is most noticeable in schools and on college campuses, where young, social people live, work, and move around in tight quarters.

This theory turns up on many websites as an explanation for the seasonality of the flu, providing yet another reason (as if you needed one) to be cautious about believing what you read on the internet. It sounds like a neat explanation, but when you dig deeper you find many problems. For most adults in the West, the amount of time we spend indoors with others does not vary across the seasons. We go to work year-round and, other than a lunch break outside if it is warm enough, we don't vary the amount of mingling we do. While students return to school in August or early September, they don't start feeling achy and feverish until November, with the rest of the population. In the summer, we tend to use public transportation *more* than in the winter, and while doing so are more likely to be sneezed or coughed on. Yet there are very few cases of influenza seen in the summer. It's confounding. "If close contact were all, one would think that London Transport would ensure an all-the-year epidemic," wrote the British virologist Christopher Andrews, who was part of the team that first identified human influenza A in 1933. Cruise liners operate year-round, and despite the close contact of passengers, the pattern of influenza aboard these floating cities follows that on land.

The British astrophysicist Fred Hoyle theorized that the flu is connected to sunspots, which are magnetic flare-ups that discolor the surface of the sun. To be sure, Hoyle was a rather controversial theorist. He once suggested that viruses and bacteria did not evolve on Earth but rather arrived by comet, as microscopic alien hitchhikers. Hoyle dismissed the widely accepted Big Bang theory and believed instead in a steady-state universe that had always existed. So he was not out of character when he proposed that pandemic influenza had something to do with solar activity. In 1990 he published a letter in the prestigious scientific journal *Nature*, in which he pointed to a relationship between sunspot activity and influenza outbreaks. He speculated that the two might be linked. Hoyle noted that a recent flu epidemic in Britain aligned with one of the biggest sunspot flare-ups on record, and provided a graph showing the relationship between worldwide pandemics and sunspot activity. Each peak in the sunspot cycle was accompanied by an influenza pandemic. Hoyle reasoned that the intense electrical flares pulsing from the sun into our orbit could drive charged particles of the virus down from the upper atmosphere and into our noses.

At this point, you're probably rolling your eyes. But let's take a deep breath in the name of science. Increased solar activity does indeed have an effect here on the ground. If the sun's activity could, in the words of NASA, "blow out transformers in power grids," shouldn't Hoyle's suggestion at least be entertained? Since we can measure the cycles of increased solar activity, should we include these in an explanation of the seasonality of the flu? The big problem with the sunspot theory is that the definition of a pandemic is so subjective that it can be manipulated to fit any model or argument. It is therefore not surprising that Hoyle's theory remains on the fringes. Instead, epidemiologists are less concerned with the sun's *spots* as an explanation of flu's seasonality, and more focused on the sun's *light*—and the way it controls our levels of vitamin D.

The vitamin D theory has to do with a loss of our immune function over the winter months. In winter in the Northern Hemisphere, the sun is at a lower angle in the sky, resulting in fewer hours of daylight. This leads to less melatonin and less vitamin D production, which leads to immune suppression. That makes us more prone to illness, and more likely to catch the flu. In other words, flu epidemics may have something to do with the length of the day and our exposure to the sun's light.

Vitamin D plays a key role in our health. While you can get some of it from your diet, most of your vitamin D comes from sunlight. After we manufacture a kind of cholesterol called 7-dehydrocholesterol, it is transported to the skin, where the sun's ultraviolet rays transform it into vitamin D. Vitamin D primes our white blood cells for action against invading microbes. Some of these white cells, called macrophages and natural killer cells, release peptides and cytokines into the cells that are infected with the flu virus or with the bacteria that follow. Without vitamin D, these white cells, which are the backbone of our immune system, don't work very well. Indeed, they might not work at all. And without our natural killer cells doing their natural killing, we become vulnerable to all kinds of viral and bacterial diseases.

What happens in places where there is significantly less sunlight in the winter? I grew up in London, where the gloomy winter sunrise might be as late as eight a.m. and sunset as early as four p.m. Trudging to and from school in the dark was not just depressing—it was also dangerous for my immune system. The British have lower vitamin D levels than those who live in sunnier climes. By some measures, twice as many people die over the dark winter than when the sun shines. The problem is especially severe for Britain's senior citizens. Their long-sleeved clothing, while protective against the cold, limits their exposure to whatever sunlight there is. Low vitamin D is also much more common in African Americans than it is in pasty English people like me. In fact, they are more than seven times more likely to have low levels, because the melanin in their skin reduces the ability of sunlight to convert 7-dehydrocholesterol into vitamin D. We don't know if this results in a greater incidence of influenza among African Americans, but their mortality from pneumonia and influenza is 10 percent higher than in the white population. This, too, supports the observation that vitamin D, and therefore the sun and our relationship to it, plays an important role in modifying our immune response.

Forty years ago in the Soviet Union, researchers hypothesized that Russians living in the far north were more susceptible to the influenza virus during the short days of winter than in the sun-drenched summertime. To test this they gave two groups of patients a flu vaccine that contained a weakened form of the virus. One group got the vaccine in the summer, when the days were very long, and one got it in the winter, when sunlight near the Arctic Circle is scarce. They found that the winter group was eight times more likely to develop a fever as a side effect of the vaccine. Less sunlight means less vitamin D. Which means a poorer immune system. Which means more side effects from the influenza vaccine.

Vitamin D makes for a well-functioning immune system, so could extra amounts of it in our diet actually prevent the flu entirely? Perhaps. In one experiment, a group of schoolchildren in Japan were given either vitamin D supplements or a placebo. The vitamin D group had significantly fewer cases of influenza. However, a similar study of healthy adults in New Zealand failed to show any decrease in the number of viral infections. And when older adults took extra doses of vitamin D it didn't improve their immunological response to the influenza vaccine, which is disappointing (especially for Britain's senior citizens). Clinicians faced with these kinds of conflicting studies can pool all the results together and then analyze the findings. This is called meta-analysis. One such analysis pooled the results of eleven studies of vitamin D. It showed that vitamin D was indeed effective at reducing the number of flu-like illnesses, but that it was no guarantee you could prevent them altogether. In other words, you can ingest heaps of vitamin D and still end up catching the flu.

The vitamin D theory suggests that the seasonality of influenza is due not to a feature of the virus, but to a feature of our immune systems. If we could somehow maintain our defenses against the virus year-round, we would not experience a winter uptick in flu. In contrast, some researchers believe that its seasonality can best be explained by features that have nothing to do with our immune system, or the sun itself. Instead, they blame the weather.

Jeff Shaman, an epidemiologist at Columbia University, uses computational modeling to predict the next influenza hot zone. He started out as a geophysicist, studied immunology, and later worked in climate and atmospheric science. His doctorate focused on modeling mosquito-borne disease transmission and its relationship to the weather. It was while investigating the transmission of the West Nile virus that he started to think about the far less exotic—but far more common—influenza virus, and how it is affected by humidity.

In 2007 a group from New York's Mount Sinai School of Medicine had looked at the role of cold air and humidity in the transmission of the flu virus. They used guinea pigs as their guinea pigs. These animals are very susceptible to infection by the human influenza virus. After placing cages of infected guinea pigs next to cages of uninfected ones, the researchers blew air from the former to the latter, while varying both the temperature and the humidity. They found that when the temperature and humidity were both low, there was a high rate of transmission. The virus became less transmissible, however, as the humidity and temperature increased. In fact, once the temperature reached a balmy 86 degrees Fahrenheit (or 30 degrees Celsius) there was no transmission of the influenza virus at all. The uninfected animals remained happy and healthy.

This finding led Shaman to study the effect of humidity on influenza, and to build computer simulations that predict the location of the next influenza outbreak. Scientists from the CDC were also interested in the relationship between humidity and influenza transmission. In one experiment, they created a simulated coughing machine using "a metal bellows driven by a computer-controlled linear motor." They armed their cough-bot with various sizes of flu-coated particles and aimed it at a mannequin whose mouth was replaced with a particle counter. They recorded the amount of virus that was transmitted, then changed the humidity of the room and repeated the experiment. In low-humidity environments, the virus particles remained infectious up to five times longer than in high-humidity ones. So by keeping the humidity high, it was theoretically possible to reduce the amount of influenza in the air. Practically speaking, though, using humidifiers to defeat influenza is not a viable plan. Although a segment of the population may install and use humidifiers, their overall use is very limited. Humidifiers are likely to be very low on the list of priorities for those with tight budgets. And humidifiers are less commonly used in indoor public spaces—where we do most of our coughing and sneezing at one another.

If humidity helps to explain why influenza is seasonal, and humidity is something we report in the weather forecast, could we predict future outbreaks based on the weather? There are real challenges with this because the spread of the flu depends on many factors, each of which changes in ways that are not predictable. In one academic paper on the subject, Shaman and a colleague wrote that "infectious disease dynamics are nonlinear and intrinsically chaotic." No academic journal would allow its authors to write the straightforward version: an infectious disease refuses to follow simple rules.

But even though weather forecasting is also nonlinear, we have incorporated it into our daily lives. It, too, follows very complicated rules and has shifting variables, which is why accurately forecasting even next week's weather is a very hard thing to get right. The components of a flu forecast are similar to those of a weather forecast. Instead of tracking cloud formation, we measure humidity. Instead of mapping how heat moves through the atmosphere, we look at how the flu moves through a population. Instead of radar and satellites, flu forecasters rely on the microbial equivalent: throat cultures and rapid flu tests provided by emergency departments and doctors' offices. This gives forecasters some understanding of the flu in real time. Just like a TV weather forecaster with a live radar, flu forecasters report on conditions as real-world observations become available, and they constantly recalibrate.

Storm warnings contain two or three different paths that the storm *could* take, each with its own likelihood of occurring. This is known as ensemble forecasting. The ensemble is based on dozens or hundreds of data points; each may predict slightly different outcomes, but when combined they produce a most likely and a least likely scenario, together with a few scenarios somewhere in the middle. Flu forecasters now produce ensemble forecasts of what could happen over the flu season. The final prediction consists of a few possible scenarios and how likely they are to occur.

Shaman and his colleagues used their forecasting model during the flu season that began in the fall of 2012. They estimated the spread of flu in 108 cities in the United States and produced weekly real-time forecasts. At first, the forecasts didn't seem to be useful. Their overall accuracy was so low that if they'd been given by a weather forecaster, you would have turned to another channel. But the model rapidly improved as they added more data from the field. By the end of the season their weather model was about 75 percent accurate in its reflection of the flu. That's not perfect, but it's far better than predictions based solely on historical data.

Shaman's flu forecasting success that year caught the eye of the CDC in Atlanta. In 2014 they declared him the winner of their Predict the Influenza Season Challenge—an accolade that came with a prize of $75,000. Based on his successes, Shaman has big dreams for the future of flu forecasting. During the flu season, he would like to see a nightly flu forecast alongside the familiar weather forecast we already get on the ten o'clock news. This isn't as odd as it sounds. After all, pollen counts and pollution warnings have long been part of our TV weather reports.

As an ER doctor, I'm not sure what I could do with a flu forecast. If the weather forecast predicted an 80 percent chance of rain, I would leave home carrying an umbrella. But what on earth would I do with a flu forecast that predicted with 80 percent certainty that the flu season would peak in a week?

With an accurate flu forecast, Shaman hopes that hospitals would change their staffing patterns and, if things look really bad, prepare extra equipment like ventilators. In a normal flu season, hospitals can certainly cope with the few additional patients who have respiratory failure due to a complication of influenza. But in a pandemic on the scale of the 1918 outbreak, an enormous number of patients would need help.

As an example, take metropolitan Atlanta and a hypothetical flu pandemic lasting eight weeks. At the peak of the outbreak, an estimated 2,000 patients a week would require admission to the hospital and more than *three-quarters* of the beds in the city's intensive care units would be occupied by patients with influenza. Almost half of the existing ventilators would be dedicated to keeping the sickest of them alive—that's in addition to the ventilators needed for patients who are in the ICU for other reasons. This is where Shaman's flu forecasting might be very useful. Because it estimates the number of likely cases *and* when the pandemic will peak, it would give hospital administrators and public health officials time to plan ahead.

It's a great idea, but I doubt it would work, because of the realities of hospital management. I've been in emergency departments for years but have yet to see any of them change their staffing patterns or medication supplies in response to a bad flu outbreak. Hospitals rarely take these steps, which are either costly, impractical, or both. Whose surgery should they cancel to make a bed available in the intensive care unit for a flu patient who might never arrive? Many emergency departments are already operating at maximum capacity and hospitals are short of nurses. There is not much wiggle room to make more beds available. A hospital bed is like a plane seat: it generates no revenue if it is not filled. That drives hospitals to be at or near 100 percent bed capacity. Asking a hospital to keep beds open and staffed for a flu epidemic that hasn't yet arrived is like asking an airline to keep ten rows of seats unoccupied for standby passengers who may never show.

When Shaman discussed his flu forecast with public health officials, they were skeptical. They suggested that he forget the whole idea and instead just encourage people to get vaccinated. But Shaman believes his flu forecast can target vulnerable populations and help increase the rate of vaccination in the United States, which is around 40 percent in adults. Once you receive the vaccine, your immune system takes a couple of weeks to mount an adequate response. Timing is essential. Public health campaigns that encourage immunization might be improved if they were based on the actual risk of influenza for that year.

Here, too, there are lessons to be learned from one kind of rather severe weather forecast: hurricane prediction. Communities exposed to warnings that incorrectly predict a hurricane change how they respond to these warnings in the future. They are less likely to listen to them.

Public health advisories about the flu would be more effective if they were coupled with an accurate forecast of how bad the season will be. The message would not simply be "Get vaccinated." Instead it would be "Get vaccinated now in time for the peak in flu cases, which is predicted to occur in ten days."

The seasonality of the flu is less of a mystery than it once was thanks to the work of Jeff Shaman and others. Humidity, sunlight, and temperature each seem to play a role, but as anyone working in the field will tell you, there is more waiting to be discovered. Our ability to accurately forecast the flu's peaks and troughs seems to be improving, but a daily flu forecast is not around the corner. Perhaps the best way to battle influenza is to forget about when it will make its move and head it off at the pass, to prevent its spread in the first place. Drugs exist to do just that, and they are so important that they're kept in secret stockpiles. They are prized and protected—and enormously profitable for their makers—but do they revolutionize our fight against the flu, or merely act as a security blanket that reassures rather than cures?

8

THE FAULT IN OUR STOCKPILES: TAMIFLU AND THE CURE THAT WASN'T THERE

Across the United States there are a handful of anonymous, cavernous warehouses. They exist partly because we don't want a repeat of 1918. Packed inside are the weapons we'd need to fight another global pandemic. The bunkers contain our Strategic National Stockpile of emergency medicines, and they symbolize both the strides and the limits of our medical prowess and logistical readiness.

Apparently, each warehouse is the size of a couple of Walmarts joined together. The stockpile was created by the CDC in 1999 and now has an inventory of medicines and emergency materials that cost more than $7 billion. The stockpile, which is maintained by the CDC, contains supplies for a public health scenario that overwhelms state and local government. It has antibiotics and vaccines and antiviral medications, surgical equipment and life support machines, and protective suits like the kind we saw during the Ebola outbreak. Its inventory is massive, and the details are classified. Originally, the goal of the stockpile was to provide medical supplies in the event of a chemical or nuclear attack, but recently the inventory has expanded to an "all hazards approach" that includes hurricanes and earthquakes. And, of course, a flu pandemic.

Maintaining the stockpile is an enormous undertaking. Medicines, just like cartons of milk, have expiration dates. While your local super-market makes sure to bring its soon-to-expire products to the front of the shelves, medicines in the stockpile just sit around waiting for a massive crisis. Once they expire, they must be replaced. Keeping the whole thing going costs at least $500 million per year—which you might consider a small price to pay for peace of mind.

The Supply Service Center, a warehouse in Perry Point, Maryland, doesn't have a name as grandiose as Strategic National Stockpile, but it plays an important part in the year-round federal response to influenza. Perry Point looks pretty, even in the late fall as the grass turns brown and the trees become branchy skeletons. The center sits on a peninsula minutes from the Delaware border, where the Susquehanna funnels into the Chesapeake Bay, a couple of hours' drive north of Washington, D.C. Perry Point is like a quaint little town all to itself, with a baseball diamond, a community center, and several cul-de-sacs, all bordered by forest and fresh water. The veterans' hospital that shares the peninsula has its own police and fire department, and even its own post office. Mark Burchess is the deputy director of Supply Chain Management Services, which is part of the Department of Health and Human Services.

"A cross between Amazon and a local pharmacy," Burchess calls his workplace, which is a warehouse that serves two purposes. The first is to distribute medical supplies to those who work in the federal government. Burchess has everything, or so it seems: vaccines for an embassy, disposable gloves and masks for first responders, gas sensor "sniffers" to detect biological threats, and blankets for hurricane victims. If he doesn't have it in stock, he knows where it can be purchased. The second purpose of the warehouse is to maintain "stockpiles of national significance" like those that would be needed during a pandemic flu.

Clients from across the federal government store supplies in the warehouse the way you might store stuff in your basement. There are boxes for the Department of Defense and pallets for the Department of

Homeland Security, as well as material stored by BARDA, the Biomedical Advanced Research and Development Authority. BARDA is a government office that develops and purchases drugs and equipment that would be needed in a public health emergency.

Burchess was excited that the center, working with BARDA, was about to double the size of its refrigeration storage unit. He couldn't say what it would hold, other than "stuff" that was vital to an influenza response. Precautions were plenty. The refrigeration unit came with two standard compressors, a third compressor for backup, and enough spare parts on-site to build a fourth compressor, should the need arise. The center also had enough gasoline on-site to run the electrical generators for more than a week. The warehouse would soon acquire medications for multidrug-resistant tuberculosis that it would store on behalf of the CDC. This place was one huge doomsday prepper campus.

On the soaring shelves that seem to stretch on forever are medicines, antidotes, and colossal boxes containing vials of flu vaccine. Everything here seems to come in wholesale sizes, as if we were at Costco. Vaccines are stored in handy five-gallon drums, each containing thousands of individual doses. The strain is printed clearly on the side of each drum. Many names would be familiar to most physicians, but others are mysterious or classified.

From Perry Point, the center sends supplies across the world every day, from autoclaves that sterilize instruments to massive electrical generators. Although owned and operated by the U.S. government, it does not receive government funding. Instead, it works as a business, making money on the items that it ships. If the CDC needs ten thousand syringes for a vaccine trial in Africa, it turns to Burchess, who gives a price that includes a markup to run the business. Like other businesses, he has competition. Several other government agencies, like NASA and the Department of the Interior, have similar supply centers, so prices remain competitive.

Every year the Supply Service Center ships out the influenza vaccine, as well as the paraphernalia needed to run a successful vaccination program: needles and syringes and gloves and inoculation cards. Burchess is especially proud of the vaccines he shipped to Customs and Border Protection during the 2009 H1N1 influenza outbreak, helping protect those who protect us. During that outbreak, almost 120 million doses of the H1N1 vaccine were shipped across the United States. The center not only distributes influenza vaccines but, rather surprisingly, collects them too. Many of those 120 million doses were given, but millions were left unused in hospitals, doctors' offices, and warehouses around the country. Each state dealt with the unused doses differently, and some declared them medical waste. What was delivered by FedEx one day might need to be picked up by a team in white protective suits the next, and the center provided a reverse logistics operation to bring many of the doses back to a central location for destruction.

In 2009 Perry Point also held most of the U.S. stock of the anti-influenza drug peramivir, which at the time did not have approval from the FDA. As a result, it was being provided under an emergency-use authorization. The team at Perry Point made sure that it was delivered to patients within twenty-four hours of the CDC's request. Peramivir, sold under the trade name Rapivab, is one of three medications that belongs to a class of drugs called neuraminidase inhibitors. The other two are zanamivir, whose brand name is Relenza, and oseltamivir, branded as Tamiflu.

"Neuraminidase" is a clunky word, but what this viral enzyme does is quite elegant. Once the flu virus gets into your cells through the cell membrane, it takes advantage of the cozy conditions to replicate. Then the baby flu virus particles have to escape the cell. They make their way to the surface and then push through the cell membrane. At first they remain tethered to the cell membrane like dinghies attached by ropes to their mother ship. Neuraminidase is the enzyme that acts like a knife. It allows the baby flu viruses to cut the rope and make their own way out and into the world. Without neuraminidase, baby flu particles could not spread the infection and replicate. The neuraminidase inhibitor drugs prevent viral neuraminidase from working. No knives, no free-ranging influenza viruses.

Neuraminidase inhibitors were first discovered in the 1960s by a group of researchers from Scotland, but it wasn't until thirty years later that scientists began to test them. The drug was first manufactured as a powder that needed to be inhaled, and after a few clever tweaks an oral formula was developed. That first drug was oseltamivir, marketed as Tamiflu. It was promoted as a silver bullet against influenza and a powerful weapon in our strategic stockpile. Rapivab is only available intravenously, so it is prescribed far less often than the other two drugs, and only to very sick patients. Relenza, the third drug, needs to be inhaled, and it captured much less of the market than Tamiflu, the best-known neuraminidase inhibitor.

In 2014 a TV ad for Tamiflu ran more than eleven thousand times. "Prescription Tamiflu attacks the flu virus at its source," said the voice-over. "Sometimes, what we suffer from is bigger than we think. . . . The flu is a really big deal, so why treat it like it's a little cold? Treat it with Tamiflu."

Another ad targeted mothers whose children had the flu. "The flu virus. It's a really big deal," a narrator said as a worried mom watched over her coughing daughter. "Mom knows it needs a big solution: an antiviral,"

But given the way the medicine actually performs in patients, it turns out that it is not much of a solution at all. If you read the small print accompanying the ad carefully, it claims that on average, Tamiflu shortens a child's flu symptoms by about a day—but only if taken within the first forty-eight hours after the onset of symptoms. After that, it's even less effective. The helpful narration—mom and her healthy daughter are now shopping for fresh fruit—tells us that children and adolescents taking the drug "may be at an increased risk of seizures, confusion, or abnormal behavior," but you shouldn't worry because "the most common side effects are mild to moderate nausea and vomiting." The drug package itself notes that vomiting occurs twice as often in children taking Tamiflu compared to those not treated. In fact, Tamiflu often gives you some of the very symptoms you are trying to relieve, and at best will shorten your misery from influenza by a day.

But Tamiflu is part of the Strategic National Stockpile. There are clearly experts who are convinced of its benefits. The fascinating story of Tamiflu played out over several years and across multiple continents, and it shows how decisions made more than two decades ago still influence our approach to the treatment of influenza today.

The story begins with the outbreak of H5N1 avian flu in Hong Kong in 1997. The virus infected only eighteen people, but six of them died. This got the attention of the World Health Organization. Two years after the outbreak the WHO published a report that mentioned that "two closely related compounds have been developed that bind to the active site in a minor protein found on the surface of influenza viruses, the enzyme neuraminidase." Trials of these compounds were under way, and if they were approved for use, they might be useful in the treatment of influenza, regardless of the specific strain. The WHO report disclosed that it had been written "in collaboration with the European Scientific Working Group on Influenza (ESWI)." But it turned out that the ESWI was funded by at least *seven* pharmaceutical companies that stood to profit from an outbreak, or at least from the fear of one. These manufacturers joined together to promote "a favorable climate" in Europe for vaccines and related research. Several years after the report came out, it was revealed that one of its authors was a paid consultant for one of the drug manufacturers. Both the ESWI and the WHO were supposed to provide objective scientific advice, but there were clear conflicts of interest. The push toward stockpiling Tamiflu began in earnest on the basis of a specious and probably biased recommendation.

Around the time of the 1997 avian influenza outbreak, neuraminidase inhibitors were in the early phase of clinical trials, but there was still enough data to make some recommendations regarding their use. In 1999 the Cochrane Collaboration, an international group of thirty-seven thousand medical contributors, issued the first of three reports about neuraminidase inhibitors. When it comes to how well drugs work, Cochrane reviewers pledge to issue reports that are "free from commercial sponsorship and other conflicts of interest." They don't just take the manufacturer's word. They look for any published and unpublished papers, and all the trials that were reported or not reported. Using all this data they make recommendations about everything from vitamin E during pregnancy (it doesn't work to prevent early labor) to yoga for back pain (seems to be better than no regular exercise). I've used their work to help me make decisions about which drugs to recommend to my patients. Cochrane found that as a *treatment* the neuraminidase inhibitors only shortened the duration of flu symptoms by about one day, although they were somewhat more effective when used to *prevent* influenza.

In late 1999 the FDA approved Tamiflu as a treatment for influenza, but within a few months it sent the manufacturer, Hoffman–La Roche, a warning letter about its advertising campaign. The FDA found that the Roche campaign lacked balance, contained misleading information about how the drug worked, and overstated claims about the drug's effectiveness. Roche had claimed that Tamiflu "reduces the duration of flu so you can feel better faster," but this blurred and exaggerated the evidence from clinical trials. Despite these concerns, the FDA had approved the use of Tamiflu to prevent influenza, and to treat influenza in children under four years of age. In 2002 the European Union also approved the drug, in time for another outbreak of avian flu in Asia two years later.

In November 2005 the avian influenza scares prompted President George W. Bush to make the short trip from the White House to the National Institutes of Health in Bethesda, Maryland. His speech there was ominous. "At this moment," he said, "there is no pandemic influenza in the United States or the world. But if history is our guide, there is reason to be concerned."

Those were chilling words, and the president outlined how his administration planned to tackle those concerns. He described the small outbreaks of avian flu in 1997 and 2003. He told the audience he'd gotten his flu shot. And he urged vigilance in a manner similar to his speeches on terrorism.

"If the virus were to develop the capacity for sustained human-to-human transmission, it could spread quickly across the globe," Bush said. "Our country has been given fair warning of this danger to our homeland—and time to prepare."

The president presented a three-pronged approach: First there would be an effort to detect flu outbreaks earlier. Second, the government would stockpile vaccines and antiviral drugs, and request $1.2 billion from Congress to purchase enough avian influenza vaccine to immunize 20 million people. Third, the president asked the country to develop emergency pandemic plans in all fifty states and every local community. This three-part approach, said the president, would "give our citizens some peace of mind knowing that our nation is ready to act at the first sign of danger." Overall, the president requested $7.1 billion to fund his strategy, although a year later Congress approved only half of that amount. Now avian influenza was on everyone's mind, even though the chances of catching it in the United States were very low.

The 2004–2005 influenza season in the U.S. was not particularly bad, and yet there was a rapid rise in the sales of Tamiflu and other antiviral medications. In the fall of 2005 five times as much Tamiflu was being prescribed in the U.S. than had been the year before. This increase was even higher in those who had no chronic medical conditions, as well as in children. This suggests that healthy people were buying up Tamiflu, preparing for a potential outbreak. The increase had nothing to do with the amount of flu that was actually out there, since there was no more than usual. What *was* unusual was the presidential address on the matter, and the media's coverage of and hand-wringing over avian influenza. In Canada there was an even greater hoarding of Tamiflu by worried citizens. Prescriptions for the drug there rose tenfold, and threats of a shortage prompted Roche Canada to restrict distribution of the drug.

In the midst of the avian flu scare, Britain's minister of health made a somber assessment. "We must assume we will be unable to prevent it reaching the UK," said Chief Medical Officer Sir Liam Donaldson. "When it does, its impact will be severe in the number of illnesses and the disruption to everyday life."

The BBC, not known for its sensationalism, reported that unless millions of doses of Tamiflu were stockpiled, an outbreak of avian influenza in the United Kingdom could kill more than 50,000 people. And so the British government made plans to purchase and stockpile over 14 million doses of Tamiflu. In the U.S., Tamiflu was already part of the national stockpile, but the supply was more modest—only 2.3 million doses were on hand, though more was on order.

Barely two months after President Bush's speech and the news that the U.S. would stockpile Tamiflu, the Cochrane group published another analysis of antiflu medications. The authors reviewed more than thirty clinical trials of the older antiflu drugs and nine trials of the newer Tamiflu. The older drugs were no longer effective, and Tamiflu, the new kid on the block, had some serious limitations. There was no role for the drug in patients who had an influenza-like illness that was not specifically caused by the influenza virus. More startling was the lack of evidence that it could even fight avian influenza, the very disease for which it had been stockpiled. Tamiflu didn't reduce deaths from avian flu when it was used in Southeast Asia, and now the virus was showing resistance to the medication. Over time Tamiflu succeeded in one respect: it made some types of flu resistant to it. In Europe at least 14 percent of the circulating influenza viruses were resistant to Tamiflu by 2008. In short, the drug worked only to strengthen the virus that it claimed to combat.

The drug was next tested in 2009, when there was an outbreak of swine flu in the U.S. This was similar to the swine flu that jumped from pigs to humans in 1976. This flu, you may recall, was identified as an H1N1 type, a combination of swine flu viruses that usually infected pigs in America and Europe. By June 2009 there were more than 30,000 cases of swine flu in seventy-four countries. The World Health Organization declared a pandemic. The CDC hosted dramatic press conferences in Atlanta, but this H1N1 swine flu, though new, was not as virulent as experts feared. Swine flu viruses generally cause only mild disease, and were no more harmful than older, standard strains of influenza.

But the swine flu was getting a lot of media attention. Just as the nickname "Spanish flu" had made the 1918 epidemic seem exotic and alien in newspaper headlines, "swine flu" sounded threatening and feral, which contributed to another run on Tamiflu. In Boston, the law firm of Ropes & Gray went the extra mile for its 1,900 employees and their families. It gave them prescriptions for Tamiflu, no doctor's visit required. The firm reminded its staff to take the medication only at the onset of flu symptoms, but it did not mention that the antiviral was unlikely to be of any significant help. In an editorial, the *Boston Globe* admonished Ropes & Gray for being part of the problem. "While there has been very little resistance to Tamiflu among swine flu patients so far," the editorial said, "that could change as cases mount up and more physicians prescribe the drug." The CDC issued a curt statement: "This is something we would not like to see widely practiced by employers."

Leaving aside the question of whether Tamiflu even worked, there was another issue: fairness. News of the Ropes & Gray decision suggested that the medicine was not being distributed justly. A privileged minority— lawyers for global firms and those with the right connections—were at the front of the line. They had access to antiviral medications before they fell ill. The have-nots would have to wait and see. Dr. Karen Victor, an internist at a Boston hospital, said the main issue was access. "[The firm believes] Ropes & Gray's employees' appearance at work is so important," she said, "that they will put that above fairness to society."

Many countries reported that the flu virus was 100 percent resistant to Tamiflu. But by 2009, despite all the evidence that it worked poorly, Tamiflu and other neuraminidase inhibitors had become part of the national stockpile of the U.S., Britain, and at least ninety-four other countries around the world. The medication was a huge financial success. Governments ordered more than $3 billion worth of antivirals. Gilead, the pharmaceutical company that initially developed Tamiflu, reported royalties of over $52 million in the first quarter of 2009. And Roche, the Swiss giant that licensed the drug from Gilead, made even more: $590 million in sales in one quarter alone. And then, rather suddenly, academics struck another blow against Tamiflu.

Back in 2003 a group of researchers led by Dr. Laurent Kaiser in Geneva had looked for any papers that examined how Tamiflu performed, and then pooled the results. They'd found ten studies, though only two had been published. The other eight had been reported on to some degree, or not reported on at all, and were otherwise languishing in the drawers (or on the computers) of their authors. The review by Dr. Kaiser and her colleagues was funded by Roche, the manufacturer of Tamiflu, and it concluded that the medication reduced lung complications, antibiotic use, and hospitalizations in otherwise healthy patients. It was a very important paper for two reasons. First, it provided evidence that Tamiflu worked, right around the time when there was discussion about placing it in the national stockpile. And second, it was used as evidence in later reports by the prestigious Cochrane Collaboration.

But in July 2009 Dr. Keiji Hayashi, an astute Japanese pediatrician, contacted the Cochrane group with a concern. He wondered what was in those eight unpublished studies, and why their findings had not been shared. Without knowing the results of the unpublished Tamiflu studies, the Cochrane reviewers might find only studies that cast the drug in a good light. The Cochrane group had indeed relied upon the Kaiser paper, which itself had relied on those eight unpublished studies. They had erred in doing so. They acknowledged this error and spent the next several years trying to fix the mistake.

Why were eight of those ten trials about Tamiflu never published? Perhaps they were poorly conducted or too small in scope, and would therefore not be rigorous enough for a scientific journal. But knowing the details of a study is a critical part of the scientific process. For example, if the effects of Tamiflu had been studied only in young and otherwise healthy patients, we could not reach a conclusion about how the drug might work in those who were elderly, or those with other medical conditions. These details are critical in evaluating any clinical trial. Kaiser's group reached its conclusions based on evidence that it alone had seen, and that no one else could properly corroborate. That's bad form, to say the least.

It is also possible that the unpublished trials actually found that Tamiflu had no benefit. Medical journals are a business and, like drug manufacturers, they have a bottom line. They're more likely to publish a positive and exciting study that will make headlines. Negative trials (in which it turned out that a new drug didn't work as claimed) are rarely submitted for publication (that's the fault of the researchers), and if they are, they're often rejected by journal editors as not exciting enough to be published (that's the fault of the editors). And negative trials usually end up at the bottom of an electronic filing cabinet. But negative drug studies are crucial if the goal is to properly judge the safety and effectiveness of a treatment. When there's a bias inherent in the publication process, we can't be sure that *all* studies, both positive and negative, have been aired publicly. If doctors don't have access to all the data, they can't be sure that they are making the best scientific regulations.

After the problem was raised by Dr. Hayashi, the Cochrane group, led by Dr. Tom Jefferson, got to work. Jefferson contacted Roche and asked the company to provide the missing data. At first, the company refused to release the data because, it said, another team was already performing a review of it. When Jefferson asked why that should prevent the Cochrane group from also reviewing the data, Roche agreed to release it—but only if Jefferson signed a confidentiality agreement. That would prohibit him not only from sharing the data but also from acknowledging that he had even signed the confidentiality agreement itself. Jefferson never signed, but Roche did agree to release some of its Tamiflu data. When Jefferson reviewed it, there were too many missing details to reach any conclusions. In December 2009 Jefferson released an updated review without including any of the unpublished data. They summarized that neuraminidase inhibitors, including Tamiflu, have "modest benefit—reduction of illness by about one day." The review concluded that "they should not be used in routine control of seasonal influenza." In addition, they questioned the effects of Tamiflu on lower respiratory tract infections. And because these drugs did not prevent infection or stop nasal viral excretion, "they may be a suboptimal means of interrupting viral spread in a pandemic."

The Tamiflu controversy then broke from the ivory tower of academia and swept into the British Houses of Parliament. There, a member named Paul Flynn sponsored a motion that was polite in phrasing but excoriating in message. It used words that would make any drug manufacturer blanch: "surprised," "uncertain," "concerned" and, most pointedly, "fatal side-effects including heart attacks." "It is unwise to continue with . . . the programme," the motion concluded. Outside of Parliament, Flynn went further by suggesting on his blog that the leftover stockpiles of Tamiflu could actually be used for something practical: salting Britain's snowy roads.

Flynn authored a report on behalf of the European Parliamentary Assembly, which found a general lack of transparency in the way the swine flu pandemic had been handled by the World Health Organization. These criticisms were echoed in the *British Medical Journal*, which published a series of articles questioning the industry ties of several experts at the WHO. Dr. Fiona Godlee, the editor in chief of the journal, noted that the WHO's guidance on the use of antivirals in a pandemic was written by a flu expert who was getting paid by the manufacturer of Tamiflu. That hardly inspired confidence in its findings.

Over the years, industry and the medical community continued their tug-of-war. Roche refused to release its internal data to investigators. Then, feeling the heat, it commissioned an independent review of Tamiflu, in which neither the authors nor their institution received any funding for their work. Roche statisticians cooperated with the review, answering any data-related questions that arose. When it was published in 2011, the review showed that Tamiflu reduced some of the complications from influenza that would have required antibiotics. This might be useful in a pandemic and could, if verified, provide a limited use for the drug. In order to shake loose the truth, the *British Medical Journal* began to publish correspondence between the Cochrane group, Roche, the CDC, and the WHO. It was all part of the journal's "open data" campaign. Sunlight was proving to be a better treatment than Tamiflu.

Within a few months, Roche released all the trials requested by the Cochrane group, as did GlaxoSmithKline, which produced another neuraminidase inhibitor, zanamivir. This allowed the Cochrane group to complete an updated review, which it released in April 2014. Finally Cochrane could produce an analysis based on all the published and unpublished clinical trials of Tamiflu (and its cousin zanamivir, marketed as Relenza). Cochrane found that, when taken as prevention, these drugs might reduce the risk of developing influenza; once a patient was sick, they reduced the duration of symptoms of the flu by less than one day. However, Tamiflu was most efficient at triggering side effects: nausea and vomiting, and sometimes psychiatric effects like hallucinations, anxiety, and even seizures. It could also damage your kidneys. The most damning finding was that Tamiflu did not reduce the risk of hospital-ization, or of pneumonia, which was the very reason the United States and other countries had put Tamiflu in their strategic stockpiles. The Cochrane report, and several others that followed, blew a hole in the notion that Tamiflu should be stocked at all.

The debate over Tamiflu continues today, although perhaps with less intensity. New academic papers have been published, but they seem to have done little to change anyone's mind. In January 2015 the *Lancet* released an analysis that included all the published and unpublished Roche-sponsored trials, and any other relevant trials reviewers could find. They found that Tamiflu decreased the risk of hospitalization and confirmed the already well-documented finding that Tamiflu shortened sickness by about a day. The study was not supported by Roche directly but by a foundation called the Multiparty Group for Advice on Science, which is (you guessed it) supported by Roche.

Despite all the controversy, and the corruption of the study pool by pharmaceutical interests, Tamiflu continues to be tested in clinical trials. As of 2017 there were at least eight open trials in the U.S. and Canada focused only on groups that are at high risk for the flu: the elderly, those who have underlying lung or heart conditions, or those whose immune systems are not working properly. For the rest of us, taking Tamiflu to prevent or treat influenza is a waste of time. Even the CDC implies this. Its most recent guidelines recommended antiviral drugs only for those high-risk patients.

But Tom Frieden, who was director of the CDC in 2014, had strongly supported the use of antiviral medications. At a telephone news brief-ing held late that year, he told listeners the opposite of what the data showed: that antiviral drugs can soften and shorten illness and reduce the risk of dying from the flu. "Antiviral treatment is particularly impor-tant this year," he said before name-checking Tamiflu, adding: "If you are sick, talk to your doctor promptly about getting antiviral treatment."

A reporter from Reuters then asked Frieden how he could square his recommendations with the evidence of Tamiflu's ineffectiveness. Frieden replied that the CDC had looked at all the data, both published and unpublished, and that there was "strong" evidence that Tamiflu was impactful. It was not, he admitted, a "miracle drug," and he wished there were better options. But there weren't. The logic, then, is to take the drug not because it's effective—it isn't—but merely because it exists.

Peter Doshi has been following the Tamiflu saga for years. He is an editor focusing on drug regulation and marketing at the *British Medical Journal*, which campaigned for greater transparency about Tamiflu. Doshi is also a member of the Cochrane group that reviewed the neuraminidase inhibitors, which makes him an authority on the issue. Doshi had studied medical history and East Asian studies at Harvard and earned his doctorate from MIT, where his thesis was on the medical politics of influenza. After his postdoctoral work at Johns Hopkins he moved across town to the University of Maryland, where he teaches in the School of Pharmacy.

We met there, where he has a commanding view of downtown Baltimore from the twelfth floor. In his uncluttered office there was a heavy copy of Wittgenstein's *Philosophical Investigations* on the table where we sat. It is a book that identifies the relationship between language and reality, and so mirrors the debate about Tamiflu and influenza treatments as a whole.

Doshi is deeply troubled by how governmental agencies had conflicting assumptions of flu treatments. Inside Tamiflu packaging is an insert that contains, in very fine print, all the information that the manufacturer must legally disclose, like dosing and side effects. All medicines have this insert, but almost no one reads them. The Tamiflu insert notes that although influenza may be complicated by serious bacterial infections, the drug "has not been shown to prevent such complications." This is what the Cochrane review found, and what the FDA required the manufacturer to disclose.

Doshi points out that when the federal government was making its pandemic influenza plans, it seems to have come to an entirely different conclusion. Like the CDC, the Department of Health and Human Services believed that Tamiflu "will be effective in decreasing risk of pneumonia, will decrease hospitalization by about half . . . and will also decrease mortality." The positive conclusions drove the "stockpile assumption," as Doshi called it, and helped get these antivirals into more widespread use. The CDC held sway over what flu medications made it into the stockpile, and the CDC thought Tamiflu worked. And so a drug whose packaging included written doubts about its efficacy was squirreled away in large quantities for a pandemic it wouldn't be able to slow.

Doshi believed that the level of evidence required for approving neuraminidase inhibitors was too low. This started in the late 1990s, when the first of these drugs was approved, and all Tamiflu had to do, Doshi said, was to "walk over that low bar." He noted that there are, even today, questions about how the drug works. In addition to inhibiting neuraminidase, it also seems to have a direct effect on the central nervous system, lowering the fever associated with infections. If this finding is correct, Tamiflu's effect on influenza-like illnesses may be no better than aspirin's.

The controversy will continue, because the conflicting data allowed for diverging conclusions. Doshi holds out for a definitive trial whose costs, while high, would be "a drop in the bucket compared to the costs of stockpiling these drugs."

Doshi's doctoral thesis was titled "Influenza: A Study of Contemporary Medical Politics," and he understands the bureaucratic realities of influenza. If there was a catastrophic flu outbreak, we would turn to the federal government to provide us with the very best medicines available.

Who, precisely, would ever take the responsibility to declare that the antiviral medications at our disposal are not useful, and that the money to keep them in the stockpile might be better spent elsewhere?

Peter Palese is a longtime professor and chair of the Department of Microbiology at the Icahn School of Medicine at Mount Sinai (formerly the Mount Sinai School of Medicine) in New York. He has been involved in virtually every aspect of research that touches the influenza virus. His lab was the first to develop the technology that allowed the 1918 virus to be rebuilt from scratch. He tested the resurrected 1918 virus on mice to determine how dangerous it was. And he was part of the group that studied why the influenza virus spreads in the winter. Palese is an author of more than four hundred research papers studying the virus, and for him there is no debate. He disagrees with the FDA-required package information that says Tamiflu hasn't been shown to reduce flu-related bacterial infections. It was all a question of timing; once there were symptoms of the flu, it was too late for the drug to work. But it is, he believes, a very good drug if given early on.

Palese actually used the resurrected 1918 virus in a study of Tamiflu. Along with others, he infected mice with the virus and then, six hours later, started treating them with Tamiflu. He found that Tamiflu was able to protect 90 percent of the mice from a lethal infection. One problem in reading the report of this experiment, however, is that it isn't clear how many mice were used. That's a really important detail because if only a few mice were used, we might not be confident of the result. It seems like a crucial experiment to verify, but because of the difficulties of experimenting with the live 1918 influenza virus, no one has done so. So all we know from this experiment is that antivirals seem to protect some mice, when given early enough. That's something to cling to, some hope amid the depressing news about Tamiflu. But it's not much.

Palese is a fervent believer in the science and benefits of Tamiflu and equates its critics with discredited antivaxxers. To be clear, the two issues are in no way comparable. There is no link between autism and vaccines. Study after study has shown this. Meanwhile, the evidence shows that taking Tamiflu when you have influenza only reduces your symptoms by about one day. To conflate the two issues is unfair. But Palese's opinion reveals the very deep feelings that drive the issue.

A godsend, or a grift? Or something in between? The debates over the neuraminidase inhibitors is perfectly illustrated in two pieces of mail I received during the same week while writing this book. One was an information card, sent by the manufacturer to emergency physicians like me, encouraging us to prescribe peramivir, the intravenous drug for influenza. The other was my monthly copy of *Emergency Medicine News* ("the most trusted news source in emergency medicine"). On the front page was an article that announced, in a large and bold caption, that Tamiflu "belongs in the dust bin of clinical practice."

If an emergency physician can feel confused about flu medications, then patients surely may feel the same. I'm duly skeptical about a drug like Tamiflu, but imagining a repeat of a 1918-type pandemic makes me think: if an influenza virus like 1918 leaped from birds or pigs, what else do we have to fight it? Noxious fumes from a factory in Falmouth? A century after the 1918 pandemic we still don't have the miracle drug for influenza. We still cling to imperfect treatments, and seek peace of mind by stockpiling something like Tamiflu, you know, just in case. Our predicament and the limits of our power are best summarized by a former CDC field officer who told me: "Tamiflu doesn't work. Now hurry up and take it."

9

THE HUNT FOR A FLU VACCINE

Vaccination, the process of infecting a healthy person with a microbe to prevent disease, dates back at least a thousand years. Some kind of inoculation was used by the Chinese as early as the tenth century, and in eighteenth-century Bengal (now India and Bangladesh), members of the Brahman caste used vaccination in their religious rituals. Every spring, priests traveled the Indian countryside to cut and scrape people across an area of skin the size of a silver coin, drawing blood and then applying a cotton ball containing smallpox and two or three drops of water from the Ganges River.

Vaccines are one of the great success stories of modern medicine. Because of them we are no longer vulnerable to smallpox or polio or measles. The flu vaccine, however, is a different story. Its effectiveness varies from patient to patient, from population to population, and from year to year. It needs to be updated each season, and even in a good year is usually no more than 50 percent effective. We may rely on it to avoid catching the flu, but its story demonstrates how far we still are from a reliable vaccine.

The start of vaccination as we think of it today is generally credited to the work of Edward Jenner, a British physician born in 1749. Jenner was a keen observer with a deep interest in the natural world, and found time for both serious study and artistic play. He investigated everything from hydrogen balloons to the life cycle of the cuckoo, wrote poetry and played the violin, but smallpox—or rather, the eradication of it—is his legacy. Because of Jenner, this virus is not on our list of worries today.

Smallpox was a vicious disease that killed more than 75 percent of those who contracted it. In the 1700s, there was one demographic, however, that seemed to be immune: milkmaids. It had been observed that in the course of their job milking cows, women came into contact with the milder bovine version of the smallpox virus, this one called cowpox. These women then became immune to the deadlier human smallpox virus. There was something in the cowpox that protected against smallpox, and in 1796 Edward Jenner famously took material from the fresh pustules on a milkmaid's hand and inserted it under the skin of a young boy named James Phipps. After a brief and mild illness, Phipps recovered completely. Jenner then infected him with scrapings from a smallpox lesion, again and again, but the boy never got sick. Jenner named this process vaccination after *vaccinae*, the Latin word for cowpox. His technique quickly spread through nineteenth-century England and beyond, saving countless people, inspiring modifications to the technique, and changing the course of history.

Jenner's smallpox vaccine was improved and modified over the next several decades, and was soon joined by others. Louis Pasteur developed several vaccines for animal diseases like chicken cholera and anthrax, but of these he is best remembered for his rabies vaccine. Rabies was a common and uniformly fatal disease in the nineteenth century. Once a victim is bitten by a rabid animal, the virus multiplies slowly and infects the brain and nervous system. Pasteur did not know of the viral cause, but this didn't really matter. He dissected and dried out the spinal cords of infected animals and then injected the remains into test animals, which then showed immunity to rabies. What Pasteur was doing was, in fact, weakening the virus, making it a Goldilocks version. It was not strong enough to kill, and it was not weak enough to be ignored by our immune system.

One hundred years ago, during the 1918 flu pandemic, there were no flu vaccines. Remember, we didn't know precisely what was causing the flu, so we couldn't manufacture a vaccine to protect us. But this didn't stop scientists and doctors from doing something, *anything*, to combat the outbreak. In 1919 Edward Rosenow from the Mayo Clinic in Rochester, Minnesota, pleaded for his colleagues to stop bickering over the cause of the flu and to focus on the opportunistic bacteria that were actually killing people. He isolated several bacteria from the sputum and lungs of flu patients in Rochester, formulated a vaccine that contained five different kinds of bacteria, and doled it out to 100,000 people. At Tufts College Medical School in Boston, Dr. Timothy Leary (whose nephew and namesake would also become a doctor and experiment with psychedelics) produced his own blended vaccine using strains from the Chelsea Naval Hospital, a nurse's nose at Carney Hospital, and the infected wards of Camp Devens. Leary mixed these samples together, grew them on plates of agar, and then sterilized the mixture. His vaccine was sent to San Francisco, where at least 10,000 people were inoculated with it.

These and other efforts gave hope to a ravaged nation. One health official at the time wrote that the greatest value of a flu vaccine was that it reduced "fluphobia." Worry and fear were as epidemic as the disease itself, and any vaccine that provided at least mental relief was welcome. There was no evidence, of course, that any of these vaccines actually worked. Today, physicians go to great lengths to be sure that vaccine trials adhere to stringent standards, but a century ago these did not exist. For example, many of the vaccine trials were conducted on survivors of the flu, after the initial epidemic had passed, meaning that the pool was tainted with a degree of immunity.

Vaccine research didn't kick into overdrive until 1933, when the flu virus was identified. Scientists could then confront the culprit itself rather than the mess in its wake. Russians led the field at first, weakening the virus by transplanting it between chicken eggs. Around one billion people in the USSR have been vaccinated using the live but weakened flu virus, which was still in use at the end of the twentieth century. Although it appeared to be successful, the live-flu vaccine was never tested in a rigorous way and it remained a constant danger. Since it used a live virus, it could cross with other strains and morph into a more virulent version.

Vaccine researchers therefore turned their attention to creating a vaccine containing what they called "inactive" strains. The virus was still grown in chick embryos, but this time it was rendered inactive by dunking it in a bath of formalin. Although a higher dose of the inactive vaccine was needed to produce an immune response, there was no concern about the virus replicating.

For the first several years the influenza vaccine contained only one strain, the influenza A virus, because, as far as anyone knew, that was the only kind of influenza out there. In 1940 influenza B was identified, which kicked off the perpetual task of calibrating vaccines to deal with multiple evolving strains. By the 1950s we had a bivalent vaccine, effective against both A and B, but the virus, as always, was outpacing us. By the late 1970s we had to make a *trivalent* vaccine to hit three strains. For the 2016 to 2017 flu season, most of the vaccine doses manufactured in the U.S. were quadrivalent. If the past is any indicator, we may soon be using pentavalent or even sextavalent vaccines. The past hundred years have been a ceaseless arms race against an enemy with whom we cannot negotiate.

The key to a good flu vaccine is matching it to the strains that are in circulation during a given season. The challenge is that it takes about six months to produce the vaccine, and so the manufacturers have to base their ingredients on some clever detective work led by the World Health Organization. There are about 110 WHO flu centers in eighty countries that receive swabs from the noses and throats of patients with influenza-like illnesses. These centers identify the flu strains that are circulating, and occasionally they will find a new one. When this happens, they send it to one of five collaborating centers, in London, Atlanta, Melbourne, Tokyo, or Beijing, for a more detailed molecular analysis. Twice a year (in February for the Northern Hemisphere and September for the Southern) the WHO convenes a meeting to collate all the information and recommend a vaccine recipe for the upcoming season. In the U.S., the Centers for Disease Control and Prevention in Atlanta provide additional domestic data and the Food and Drug Administration makes a final decision on what goes into the vaccine. The manufacturers then have about six months to get the recommended flu vaccine to market.

Because the influenza virus can mutate so quickly, nailing the exact recipe is challenging. In some seasons the match is close to perfect, but this is not always the case. If the virus drifts after the February meeting of the WHO, there will be a mismatch between vaccine and virus. The greater the mismatch, the less effective the vaccine. In a good year, we might expect the vaccine to be 50 to 60 percent effective. In the 2004–2005 flu season, that figure was only 10 percent, meaning that the vaccine was a big misfire. We also botched the 2014–2015 season, when new H3 strains hadn't been included in the vaccine. That season the vaccine was a measly 19 percent effective, compared to over 50 percent in the previous year. As I write this, we are in the middle of the 2017–2018 influenza season. So far there have been near-record numbers of hospitalizations and the vaccine appears to be less than 20 percent effective.

Even if the vaccine hits the bull's-eye, different demographics have different reactions to it. Children have a very good response to the vaccine. The situation is more complicated with elderly patients, who have weaker immune systems overall but also wield a lifetime buildup of natural immunity. After withstanding many flu seasons, their immune systems are wiser, you might say, than those of the young.

The U.S. and most other developed countries strongly recommend that the elderly receive a flu vaccine. One study compared eighteen different groups over ten influenza seasons and found that the vaccine reduced the overall winter mortality rate in seniors by an astonishing 50 percent. But CDC epidemiologists have shown that the influenza mortality rate among seniors rose *alongside* the vaccine rate, which raises questions about the urgency to vaccinate the elderly. The bottom line is this: even if the elderly are vaccinated, they are still the population most likely to die from influenza.

One way to better protect seniors is to vaccinate an entirely different demographic: schoolchildren. This notion was elegantly demonstrated in a natural experiment in Japan. From 1962 to 1987 most Japanese schoolchildren were vaccinated against influenza; at one point the vaccine was mandatory for a solid decade. The vaccination rate grew to around 85 percent, but the mandatory vaccination program was discontinued in 1994. Over the next several years, there was an increase in the number of deaths in the elderly during the flu seasons. In the U.S., where there had been no change in the vaccination policy, deaths of the elderly over the same flu seasons remained unchanged. Vaccinating one part of the population, in other words, benefits another.

Data can be interpreted in many ways, and each nation has crafted its own policy accordingly. The CDC has recommended the flu vaccine for all healthy children in the U.S. since 2008. In 2013 the United Kingdom phased in a child flu vaccination policy, in contrast to the majority of European countries. Germany provides free vaccines only to seniors, leaving parents to pay for their children. Across Europe, the childhood vaccination rate is 15 percent, compared to almost 60 percent in the U.S. If flu vaccines are indeed mankind's greatest weapon against the flu, why are they used in wildly different capacities?

*   *   *

When my colleagues gave one another the influenza vaccine at the George Washington University Hospital, we were following the advice of the CDC. When patients with influenza started to roll into the ER a few months later, I would ask if they had received a flu shot. Many had, and yet here they were in the hospital. I knew how they felt. My only visit to the ER as a patient, that year I got a nasty case of the flu, had been after I had been given the flu vaccine.

Despite the regular failure of the vaccine, Americans are bombarded every year with reminders and opportunities to get a flu shot. By the end of August pharmacies are posting signs and doctors' offices are gearing up. The vaccine is offered at many workplaces and houses of worship, and hospitals require all their health care providers to be vaccinated. Behind this effort is the CDC, which recommends the flu vaccine for everyone over the age of six months. One CDC poster that caught my attention asked, "Who needs a flu vaccine? a) You, b) You, c) You, d) All of the above" (in case you were wondering, the correct answer is *d*.) The poster reminded us that "even healthy people can get the flu, and it can be serious." The message then got more explicit: "Everyone 6 months and older should get a flu vaccine. This means you."

Recommendations about the use of vaccines in the United States are made by the Advisory Committee on Immunization Practices (ACIP), a group of more than a dozen experts with backgrounds in vaccination research, public health, and health policy. It meets three times a year to review any new evidence and provide advice and guidance to the director of the CDC about the use of vaccines. As recently as 2006 the committee recommended the flu vaccine only for those at high risk for complications of influenza and adults over the age of fifty. But a couple of years later it expanded its recommendations to include *everyone* over the age of six months. And that recommendation has remained in place ever since.

The public health campaign by the CDC to vaccinate everyone is not, however, shared by other countries. Europe and Australia recommend the vaccine only for the very young, the elderly, and those with underlying illnesses. Healthy adults are simply not targeted. It's very difficult to compare death rates from influenza across different countries because the definition of an influenza case varies, as does the way in which a country collects its own statistics. Often, deaths from viral influenza and bacterial pneumonia are listed together. It is challenging, therefore, to compare the data we have from the United States and the UK. However, in the UK, the death rate from influenza in 2014 was 0.2 per 100,000 people. And in the U.S., it was 1.4 per 100,000. That is seven times higher than in the UK, a country that vaccinates far less of its population. These numbers must be interpreted with caution, but they do at least suggest that the approach in the UK is reasonable.

How can we properly determine whether a "vaccination for *all*" program, like the one in the U.S., saves more lives and protects more people than the English "vaccination for *some*" program? We would have to do some clever clinical studies, and because of differences in health care delivery they would need to be done within the borders of one country. Perhaps for one flu season we could encourage everyone to get vaccinated, and for the next we would encourage only those at increased risk. We could compare the influenza death rates between the two groups and get our answer.

Of course, it's more complicated than that. Because the mortality rate from influenza is so low, we would need to enroll hundreds of thousands of patients in order to see if the vaccine made a difference. We would also have to ensure that those who became sick really had the influenza virus instead of a virus that causes influenza-like illness. That would require swabbing the throats of hundreds of thousands of patients and sending samples to the lab. This would be both time consuming and very expensive. An experiment like this might also be undermined by the strains of influenza that were circulating in each year. If one year's strain was more contagious or deadlier than the next, our experiment would tell us nothing.

We could instead collect evidence from small trials and look for trends. This method was used by the Cochrane Collaborative in 2014, when they reviewed all the studies that evaluated the effects of the flu vaccine in healthy American adults. It was a large undertaking; there were ninety studies that compared giving the vaccine to withholding it, and a total of 8 million patients were involved. Some of these trials may have included only a few thousand patients, which is not enough to give a definitive answer. Others might not have been trials in which people were randomly assigned to the vaccine or the placebo group. But taken together, the weaknesses of one trial might be balanced by the strengths of another.

The Cochrane review found that the effect of the influenza vaccine in healthy adults was "small." About 2.5 percent of those not vaccinated became ill, versus 1.1 percent of those who were. That's *very* small. Put another way, you would need to vaccinate seventy-one people to prevent one case of actual influenza. The vaccine did not reduce the number of working days lost, or the number of hospitalizations. So yes, the vaccine does prevent influenza in young, healthy adults, but in a very modest way. So why does the U.S. still recommend universal vaccination, while the UK does not?

As with Tamiflu, it comes down to language. The CDC describes the flu like this in a poster intended for use in doctors' offices:

> The flu may make people cough and have a sore throat and fever. They may also have a runny or stuffy nose, feel tired, have body aches, or show other signs they are not well. The flu happens every year and is more common in the fall and winter in the U.S. People of all ages can get the flu, from babies and young adults, to the elderly.

Not so bad. But then this is on the home page of the CDC's flu site:

It can cause mild to severe illness. Serious outcomes of flu infection can result in hospitalization or death. Some people, such as older people, young children, and people with certain health conditions, are at high risk of serious flu complications. The best way to prevent the flu is by getting vaccinated each year.

The CDC's approach to flu is that it is a potentially deadly disease that can be prevented with a vaccine. The British take another approach. Here is the advice about the flu from their National Health Service:

Flu is a common infectious viral illness spread by coughs and sneezes. It can be very unpleasant, but you'll usually begin to feel better within about a week. . . . [If you are an otherwise healthy adult] there's usually no need to see a doctor if you have flu-like symptoms. The best remedy is to rest at home, keep warm and drink plenty of water to avoid dehydration.

There's no mention of death as a complication. It's all very "keep calm and carry on," just like it was during the 1918 pandemic. At most, according to the British, influenza can be a bit of a nuisance:

Most people will make a full recovery and won't experience any further problems, but elderly people and people with certain long-term medical conditions are more likely to have a bad case of flu or develop a serious complication, such as a chest infection.

Is the flu a killer or an irritant? We know with certainty that each year it kills many people in both the U.S. and Britain. And we know with equal certainty that for almost all healthy people the flu is nothing but a minor annoyance. Both are correct. That's the nature of influenza. It's tricky and mysterious, causing discomfort in some of its victims and death in others. It's just that my home country and my adoptive country quantify them in different ways.

The British version of the CDC's vaccine advisory committee is called the Joint Committee on Vaccination and Immunisation (JCVI). It meets three times a year, reviews the scientific evidence, and makes recommendations to the secretary of state for health if there is a need to change the vaccination policy. Andrew Pollard, the head of the JCVI, trained as a pediatrician and now holds the seat of professor of Paediatric (that's how it is spelled over there) Infection and Immunity at the University of Oxford. Pollard is extremely cognizant of the numerous effects of the flu, but for the JCVI, the biggest measure is cost-effectiveness.

It may seem cold or callous to fixate on costs when lives are at stake, but money and resources are limited, and reckless or misdirected spending can lead to poor medical practices or greater harm. For example, spending $1 million on medications for those who've had a heart attack might save 1,000 lives each year. That same $1 million might have been spent instead on screenings for cervical cancer, which would save the lives of 60,000 women each year. What's more important: saving 1,000 lives or 60,000 lives? It often comes down to who is asking (and which disease it is that you have).

Andrew Pollard and his team at the JCVI looked at studies that measured the cost-effectiveness of the influenza vaccine. They concluded that given the vanishingly small number of young, healthy adults who become severely ill or die from the flu, it is not cost-effective to vaccinate this section of the population.

Pollard's committee measures the cost to the health system itself: how much the vaccine costs and how much it decreases the number of days that patients spend in the hospital or in the intensive care unit. They also estimate the vaccine's effect on the number of flu-related visits to doctors' offices. What they don't measure is the wider cost to society, which includes lost labor, lost wages, or the time a parent must spend taking care of a child. These, too, are burdens on society, but they do not enter into the deliberations of the JCVI. The vaccine is cost-effective to the health care system when it is given to children, the elderly, those with medical conditions, and pregnant women. It is not cost-effective when given to young, healthy adults.

In the United States, the cost-effectiveness of the vaccine is less important. What is more important is whether or not it works. This approach has resulted in another difference in vaccination policy between the U.S. and England, this time over the vaccine for chicken pox. The varicella vaccine can prevent both chicken pox and shingles, a later complication of the disease. In the U.S., the varicella vaccine is recommended for all children; the first dose is given at twelve months of age, and a booster shot four years later. In England, the vaccine is not on the list of vaccines for children (though it is recommended for those older than seventy because it can prevent painful outbreaks of shingles). In the U.S., if a vaccine has been shown to work safely, the CDC generally recommends it.

Remember that at the beginning of the 1976 flu outbreak President Gerald Ford had to choose between two perfectly sound recommendations. One was to quickly vaccinate as many people as possible, while the other was to stockpile the vaccine and wait to see whether things got worse. Ford rejected the wait-and-see approach.

"We cannot afford to take a chance with the health of our nation," you may recall him saying. "Better to err on the side of overreaction than underreaction."

This is our overriding approach to health care in the U.S. We are always ready to do more, to try the latest medications or surgical procedures, because, well, why take a chance? Compared with other western countries, we do more invasive studies of the heart for patients with chest pain, without actually improving their outcomes. We put more of our patients into the intensive care unit, even though they are, on average, less sick than their counterparts abroad. We give more chemotherapy to cancer patients near the ends of their disease, even though it improves neither the quality nor the length of their lives. We do these things because we can, because to do otherwise would be considered giving up—even if doing less would be an extremely sensible and kind decision.

Influenza is not cancer, and it is not heart disease. But our approach to it is emblematic of the way we treat other diseases. Doing more is better. If we have an unexhausted option, we exhaust it. And because many vaccines have had spectacular success in preventing and eradicating some ghastly infectious diseases, we expect the influenza vaccine to do the same. It's another high-tech solution. To most people the word "vaccine" is tantamount to a guarantee that a disease will leave you alone. It's hard to make a catchy public service announcement that reflects the subtleties therein.

"Vaccinate everyone over six months old" is the current message. It is easy to understand and easy to remember. A more accurate message is much clumsier: "Vaccinate school-aged children and pregnant women and probably the elderly (but the evidence is mixed) and those with chronic conditions, but no need to vaccinate young, healthy adults" That wouldn't really fit on a billboard. In this case, nuance may invite danger.

The quest for a better influenza vaccine continues. The holy grail would be a vaccine that covers all possible strains of influenza (so there would be no problem of mismatched vaccination) and that needs to be given only once, not every year as is now the case. Dozens of research labs across the world have worked to create this so-called universal vaccine, but so far without success. The influenza virus is just too adept at changing its disguise, remaining one step ahead of our efforts to neutralize it with a one-shot-fits-all vaccine. Although influenza is a common illness, finding an effective vaccine to prevent it remains an exceptionally challenging endeavor.

10

THE BUSINESS OF FLU

The vaccine for the 2014–2015 flu season was only 29 percent effective against the influenza A strain. A great number of elderly patients caught the flu and died from the bacterial pneumonia that followed. In England and Wales, government statisticians saw this bump in the numbers. In January 2015, for example, there were an additional 12,000 deaths compared with the previous January. There had been a steady decline in the death rate over three decades, and yet here it was, now rising by more than 5 percent. No one was quite sure what was happening. Perhaps cuts to the National Health Service were to blame. There were fewer ambulances, longer waits to be seen in emergency departments, and more shortages in hospital staff than ever before. But there were no cuts to the Scottish health service, and the death rate there increased by more than 6 percent. Across Europe, in those older than sixty-five, there were 217,000 more deaths.

It isn't surprising that the vaccine was off that year; sometimes the vaccine matches the strains in circulation, and sometimes it doesn't. What *is* surprising is the way calculating financial institutes treated the spike in deaths. Fewer elderly people meant that fewer retirement benefits were being collected, which was a boon for those who managed retirement funds. The increase in deaths freed up more than £28 billion in pension liabilities, meaning banks and managers had a giant windfall of money to invest elsewhere. The spike was almost certainly an anomaly, but the financial sector scrutinized it for any hint of a trend. There was too much money in the game to do otherwise.

"From a pension scheme perspective, this new data is still only a snapshot," said Andrew Ward, a partner at Mercer, one of the largest human resource companies in the world. But "some significant risks remain in a world where an extra year of life expectancy can add 5% to liabilities."

These are dispassionate and rather cruel calculations, but business is business. I studied for an MD, not an MBA, and medical school then did not include any classes on the business of medicine. It should have. Health care and business are fused together. Health care costs money, and money can be made in health care. Influenza arrives with the regularity of the opening bell of the New York Stock Exchange. We buy insurance against it, stockpiling Tamiflu and other medicines in those warehouses at great cost, in an effort to prevent an even costlier pandemic, the ultimate liability. Flu affects business, and business in turn affects the flu.

It has been like this for at least a century. In 1918 the elderly lived on pensions and didn't have much life insurance. Young adults, on the other hand, were more likely to have life insurance, and were also dying in greater numbers during the pandemic, meaning insurance companies stood to lose. And that's precisely what happened. In October 1918, the Equitable Life Assurance Society paid out over seven times more claims than it had done the previous year. The Metropolitan Life Insurance Company paid out $24 million more in claims than it had expected. That's $370 million in today's numbers.

The life insurance business took a hit. But there were other economic effects of the 1918 pandemic, some of which benefited those who survived. Droves of working-aged men and women fell ill and stopped working, resulting in a labor shortage. Since so many middle-aged adults died, the labor supply dwindled, and workers demanded higher wages. In states and cities that had a higher flu mortality, workers experienced a higher growth in wages. The flu's positive impact on per capita income growth was "large and robust" in the years following the pandemic, according to one economic study. What was a tragedy for one family turned into an opportunity and a better standard of living for another.

The pandemic, though, was ruinous for many businesses. In Memphis, the railway service was cut because there was no one to run the cars. In Kentucky and Tennessee, coal mining production dropped by half. In Little Rock, merchant income declined by 70 percent. New York City staggered business hours to cut down on person-to-person contact, but owners and manufacturers lodged complaints against the Board of Health, claiming it had overreached. Soon the timetable for banks, theaters, and department stores was renegotiated to allow longer opening hours.

The flu influenced economics at an individual level too. In 1918, the poor were more likely to die of the flu than the rich. Crowded living conditions were the perfect environment for the virus to spread via coughs and sneezes from person to person. Those who were socioeconomically disadvantaged were poorly nourished and more susceptible to the disease and its complications. One report found that the poor were three times more likely to die from influenza than were the "well-to-do." Those who lived in a four-bedroom apartment had a 56 percent lower mortality compared with those who lived in a one-room apartment. The wealthy were certainly not immune—the flu killed former president Grover Cleveland's sister, family members of Sherlock Holmes creator Sir Arthur Conan Doyle, and future president Donald Trump's grandfather Frederick, but then, as now, socioeconomic class was a strong predictor of health and survival.

The financial impact of the 1918 virus even reverberated into the future and the lives of those not yet born. Douglas Almond, an economist from Columbia University, analyzed three large populations from the time: those born in 1918 who were exposed to the pandemic as infants, those born in 1919 who were exposed to it in utero, and those born in 1920 who were not exposed at all. The in utero group had higher rates of physical disability, achieved a lower level of education, and earned lower incomes as adults. Compared with the other two cohorts, the in utero group were 5 percent less likely to finish high school and earned almost 10 percent less. They were also more likely to receive welfare benefits and to end up in jail. Another study found that those exposed to the virus in utero had 20 percent more heart disease by the age of sixty. The effects of pandemic influenza were felt for decades and in ways no one could have guessed.

Some individuals and businesses benefited from the pandemic. Mattresses were in high demand, since many of the sick were marooned at home and on bed rest. Drugstores were doing a brisk trade, as were undertakers. There was a sixfold increase in the cost of funerals in Philadelphia. The *Washington Post* was outraged by what it called "the ghoulish coffin trust" that was "holding the people of this city by the throat and extorting from them outrageous prices for coffins and disposal of the dead." If your trade was death, the flu was a balm for your bottom line.

The entwining of business and health during the great flu pandemic was vividly demonstrated at the Strand Theatre in New York. In October 1918, with the pandemic now plainly in the public eye, a new Charlie Chaplin movie opened. *Shoulder Arms* was a Great War comedy that took place on the battlefields of France, and audiences loved it. Perhaps they wanted a distraction, and a reason to leave their homes. The crowds were so large that the Strand extended the film's run. Harold Edel, its twenty-nine-year-old manager, took out a full-page ad in the weekly *Moving*

Picture World. Some theaters were shunned by "panic-stricken people," Edel wrote, so he wanted to congratulate those who "take their lives in their hands to see it." At the bottom of the ad, double-underlined and in a huge font, was the recommendation of the Board of Health to "AVOID CROWDS." Edel's ad continued: "New Yorkers took their life in their hands and *Packed* the Strand Theatre all week." It was splendid news for the business during a period of dark turmoil. Edel, alas, never got to see his ad in print. He died of influenza a week before it went to press.

Nearly a century later, another man reaped rewards from the flu—or at least the fear of it—and wound up dead. In 2005, at the age of only twenty, Evan Morris was hired by the pharmaceutical giant Roche to work in its Washington office as a lobbyist. Before long Morris was in charge of a team of junior lobbyists and had a $50 million budget to influence government policy. To grease the wheels and keep politicians beholden, Morris directed $3 million in political donations to both Democrats and Republicans. He became one of the most successful lobbyists in Washington. In 2007 he bought a $1.7 million home in the suburbs of Washington. His garage contained a number of Porsches.

Washington is crawling with lobbyists. There were 11,000 of them in 2016, armed with $3.1 billion to ply lawmakers and the federal government, with the goal of supporting rules and regulations that are friendly to business. The biggest industry for lobbying that year was not the energy sector, where a paltry $301 million was spent. It wasn't the defense sector; lobbyists there deployed only $128 million. The industry that spent the largest amount on lobbying was then—and is now—health care. In 2016, health care companies spent around $500 million on lobbying, and half of this sum came from the pharmaceutical industry. Evan Morris was a big player in a big field, and he enjoyed wild success because of the flu.

In the weeks prior to President George W. Bush's 2005 announcement about new policies to address pandemic influenza, Morris had hired consultants with a single mission: to stir up fear about avian flu in order to sell more Tamiflu. Whether or not this actually influenced the president's announcement, Morris was pleased with the result. The government bought more than $1 billion worth of Tamiflu for the Strategic National Stockpile. Morris continued to lobby on behalf of Roche, the manufacturer of Tamiflu, and was compensated handsomely to do so. In 2011 he paid $3.1 million in cash for a waterfront property on the Chesapeake Bay. He called it "the house that Tamiflu bought."

Then Morris's employer received an anonymous warning of his "unusual financial arrangements," according to a riveting story about him in the *Wall Street Journal*. Morris allegedly embezzled millions of dollars to fund his extravagant lifestyle, and gave illegal kickbacks to clients. On July 9, 2015, he was summoned to a meeting with the head of legal affairs at his Washington office. Morris, realizing that he was in trouble, cut the meeting short and left quickly. He bought a gun, drove to his favorite golf course, ate a steak dinner, and bought everyone a round of drinks. He walked to a fire pit a few hundred yards from the clubhouse, a bottle of expensive champagne in his hand. He texted his wife the details of his life insurance provider and financial planner. Then he shot himself.

It's impossible to say whether one man caused the United States government to buy $1 billion worth of Tamiflu. But it's clear that one man became disproportionately wealthy marketing a flu-related product. When profit is as much of an incentive as is the public good, you get a story like Evan Morris's.

Lobbying, of course, can be a force for good, especially against pandemics. Many groups over the last few decades advocated for more HIV research. As a result, 10 percent of the budget of the National Institutes of Health was eventually set aside for that purpose. That's about $3 billion each year—a huge amount, especially given that less than 1 percent of the U.S. population is infected with the virus. Activists and lobbyists brought a neglected disease to the attention of lawmakers, which resulted in more research dollars. While HIV is still a serious disease, antiviral medications have turned it from a life-threatening plague into a chronic but largely manageable condition. Those medications would not exist without people lobbying their government.

There are several sectors that gain from lobbying about pandemic influenza. Scientists want support for their research into better drugs and better vaccines. Federal agencies like the NIH and the CDC may receive more funding. Pharmaceutical companies want to make and sell vast quantities of their products. Without a government commitment to purchase millions of vaccines and antiviral medications for the stockpile, companies wouldn't invest billions to bring them to market.

We are directing money to the problem. Nearly $213 million, in the case of one particular office in the Department of Health and Human Services. Brace yourself for more acronyms. The department's Office of the Assistant Secretary for Preparedness and Response (ASPR), created in 2006 in the wake of Hurricane Katrina, is charged with preparing for and responding to natural disasters and public health emergencies. And 11 percent of its budget goes to planning for outbreaks of infectious diseases and pandemic influenza. Is that enough?

Not according to one former ASPR official. "We are better prepared than we have ever been in the past, but we are not fully prepared for a flu pandemic," he told me. He pointed out that, unlike natural disasters, epidemics have "a slow rollout." The slow but steady manner in which influenza arrives means that we often lose focus on it; federal dollars are directed to more pressing and visible disasters, like a flood or an earthquake that strikes with no warning and causes immediate and urgent humanitarian needs.

The allocation of federal resources toward preparedness is cyclical. Something bad happens, or a function of government fails, and dollars are sent to fix the problem. As time goes by without a major recurrence, a kind of preparedness fatigue takes over. New priorities are set, and money is directed to other areas—until there's another disaster, at which point the whole cycle starts again. It's either panic or neglect. As a result, monetary support for influenza preparedness varies from budget cycle to budget cycle. In 2014, ASPR received $111 million for pandemic flu preparation. A year later the budget dropped by 60 percent, to $68 million. But the news was better in 2017; the budget for pandemic influenza increased to $121 million. Such wide swings make it very challenging to plan for more than one year at a time, and make it almost impossible to fund multiyear research programs. We need to bake influenza preparedness into the health care system, so that we can focus on the bugs and not on the budget.

Spending on preparedness is only one piece of the influenza pie. The NIH is heavily invested in influenza research. Each year it spends about $220 million on research projects that cover everything from investigating how the virus evolves to creating better vaccines and the next generation of influenza medications. The NIH estimates that every one dollar it spends on research stimulates private industry to spend eight. In that respect, influenza is not a drain on our economy. It's a steroid. It supports jobs and businesses across the country, including a start-up that will test for influenza from your living room couch.

We have in-home kits to measure our cholesterol, to tell us if we are pregnant, and to detect HIV infection. But there is not a home flu detection kit—yet. A California company called Cue hopes to provide such a kit as part of a suite of medical tests so that, in the words of its infomercial, "you can hold the power of your health in your hands." The infomercial then cuts to a child of perhaps four or five having his nose swabbed. Mom inserts the swab into the Cue machine, a silver box that sits on the desk, and then a message pops up on her iPhone: FLU DE-TECTED! Mom stays calm as she presses the CONTACT PHYSICIAN link. And presto: a doctor pops up on the screen to say that he's sending a prescription to the local pharmacy. Meanwhile, an alert is sent to Dad: FLU DETECTED IN YOUR FAMILY NETWORK.

Cue's two founders first thought of home influenza testing in 2009, during the swine flu outbreak. Given the publicity and media frenzy at that time, it was inevitable that a business opportunity would present itself. Cue initially received $2 million from angel investors, and another $7.5 million in 2014.

"The business is based on the razor-and-blade model," explained one of the founders. "We don't plan on making a profit on the razor." The razor, in this case, is the sleek silver testing unit ($199). The blades are the test cartridges, which are priced at four dollars apiece but must be replenished.

In general, similar rapid influenza tests are not very sensitive. They vary greatly, and the best have a sensitivity of only 75 percent. This means that if you have influenza, the test will detect it only 75 percent of the time. Or, put another way, 25 percent of people *with* influenza will not be told by the test that they have it. That's not very reassuring, and shows how limited these tests can be.

Cue also makes a rapid vitamin D test. Might that be useful in the fight against influenza? A better question is why measure your vitamin D level at all? All you need is fifteen minutes of sunshine on your face or arms three times a week, and your body will make all the vitamin D it needs. If you are really worried, you could take a vitamin pill each day. Just to be sure. If you did that, your vitamin D levels would be perfect. And there would be no need to measure levels every day.

If you have a couple of hundred dollars to spare and you like shiny products with a Bluetooth connection, Cue might be for you. But for the majority of us, getting a message on your smartphone that your symptoms are caused by influenza is going to be of little benefit. I doubt the elderly or chronically ill would invest in technology like this, which might be just as well. When they come down with symptoms of influenza, these at-risk groups are more likely to really need medical care—which they should seek without waiting for an app to tell them to do so.

Even though flu provides business opportunities, no one wants to see a 1918-style outbreak to help the economy. In fact, the opposite is likely to occur. To understand the scope of the economic mayhem that could result from a bad influenza outbreak, a paper aptly titled "Total Economic Consequences of an Influenza Outbreak in the United States" is a good place to look. And the news isn't reassuring. The paper's three authors, all economists, built a complex model that took into account some of the largely overlooked consequences of an influenza epidemic. Consider tourism, for example. Pandemic flu would frighten off domestic and international tourists, and all the associated industries would suffer, from airlines to hotels. Travel restrictions might decrease sales at the gas pump, at movie theaters, and on public transportation. Truck drivers would be out sick, hampering the supply of everything from heating oil to potatoes. Each of these sectors would see an economic downturn. When these and many more scenarios are factored in, the estimated cost of a severe influenza outbreak in the United States is between $20 billion and $25 billion. That's about the same as the economic loss that would result from a two-week total electricity blackout in Los Angeles County.

But it's a two-way street between the flu and business. Airlines would suffer from a pandemic, but air travel might have helped cause it. Just as passenger ships brought the 1918 virus to vulnerable communities, the confined metal tubes that carry us through the skies are perfect flu incubators. We had no idea of just how big a part they played in the spread of influenza until another devastation occurred: the terrorist attacks on 9/11. There was a dramatic reduction in flights following the attacks, and a decrease in air travel persisted for some time. That year, the peak activity of the influenza virus came two weeks later than usual.

Perhaps no big business is more quintessentially American than the yearly spectacle that is the Super Bowl. But don't encourage your favorite team too much. It turns out that if your local team makes it all the way, your risk of catching influenza *back home* increases.

Charles Stoecker is a health economist at Tulane University and has taken a deep dive into the effect of the Super Bowl on influenza. Several years ago, he was attending a research conference in Houston when he heard a news report about a shortage of exotic dancers. Huge crowds of young testosterone-charged men descended on the city when it hosted the Super Bowl, and there were not enough exotic dancers to meet the demand at local strip clubs. There was a suggestion to bus these entertainers in to fill the gap in the workforce. This got Stoecker thinking about the health consequences that might occur as a result. Might it lead to more cases of sexually transmitted diseases? Stoecker quickly realized that it would be impossible for him to access the data he would need to answer that question, but a research seed had been planted.

The Super Bowl takes place every year in February, which is often a peak influenza month. How, he wondered, might participation affect influenza deaths? Stoecker had a theory: If your local team makes it to the Super Bowl, there will be more people back home watching the game at sports bars and restaurants. More fans will hang out in close contact at Super Bowl parties, sharing food and drinks. This increased social mixing would spread more influenza, which would, so the theory went, increase the number of deaths, especially in the elderly.

To test his hypothesis, Stoecker analyzed twenty-five years' worth of data on flu deaths and on Super Bowl appearances. The title of his paper was "Success *Is* Something to Sneeze At"; sending a local team to the Super Bowl, he found, causes an 18 percent increase in influenza deaths in the elderly in the team's hometown. In the years in which the Super Bowl was scheduled closer to peak flu activity, the effect was even greater; there was a *sevenfold* increase in influenza deaths back home. Even if they did not attend Super Bowl events at the same rate as the younger crowd, the elderly were at risk because the number and mobility of people carrying the virus increased.

To be sure that the effect was real and not just a statistical fluke, Stoecker analyzed influenza deaths back home in the seasons right before and right after a city sent a team to the Super Bowl. If the flu bounce was really due to the Super Bowl, the mortality rates over these periods, when the local team was not in the Super Bowl, should have stayed the same. And they did. There was no significant change in the usual deaths from influenza. As a control for his experiment, Stoecker looked at deaths from other causes, like heart attacks, cancer, accidents, and suicides. If the mixing theory was correct, it should affect only transmissible diseases like influenza, and not any of these other conditions. And when he looked at the numbers, there was no bump in local deaths from, say, cancer in the year that a city sent a team to the Super Bowl. That is precisely what you would expect if the mixing theory was really the cause of the increased influenza mortality.

The data also showed that in the cities that *hosted* the Super Bowl, rather than the ones that sent a team there, the influenza rate was unchanged. Travelers to the Super Bowl were not infecting the local population. This, too, makes sense. Cities that host the event are generally warm despite the wintry season. That's the reason they are chosen to host in the first place. But influenza likes it cold. The weather in the host cities mitigates any effect on the rates of influenza deaths that social mixing might have.

"Your Team Made the Super Bowl?" asked a headline in the *New York Times* when Stoecker's results were published. "Better Get a Flu Shot."

*   *   *

Health care experts often talk about the burden of diseases in our society, and rank them in order of how deadly they are. That's why you've heard that the number one killer is heart disease, and cancer the number two. Influenza and pneumonia, which are listed together, come in at number eight, just below diabetes but higher than kidney disease. But digging a little deeper into the effects of influenza reveals a story far more complex than a top ten list of deadly diseases. The virus infects so many aspects of society, from secret stockpiles to Super Bowls, from the global economy to average life expectancy.

The 1918 pandemic had serious economic consequences, some of which were apparent only decades later. A century after the great pandemic, influenza threatens our economy in unpredictable ways, and we have entire industries and government offices dedicated to battling it. They consume and generate millions of dollars each year, sometimes for the greater good, and sometimes in a wasteful or corrupt fashion. Our lives—not just our personal health—are interwoven with the flu, in ways that we are only beginning to understand. Just when we think we've got a handle on its behavior, the virus dodges our grasp and defies our expectations. It is an invention of nature that so far outmatches the cunning of humanity. When will our intellect outpace the ingenuity of the flu? Not soon enough.

EPILOGUE

In the prologue we met Autumn, the hardworking and healthy mother of two who was brought to death's door by influenza nearly one hundred years after the great flu pandemic. I had talked with her as I was completing my journey to understand the virus, and had one more question to address. Are we ready for the next 1918-like pandemic, the one most experts are saying is just a matter of time? Autumn's story helped me focus my thoughts for the future in three broad categories: our *knowledge* about the virus, our *response* to it, and our *preparation* for the next outbreak.

First, there has been no greater achievement in our battle against influenza than our knowledge about its cause. In 1918, as millions lay sick and dying, we had no idea about the agent of our apocalyptic destruction. It could have been bacteria. It could have been the air we breathed or a lack of sunlight. Perhaps it was something as mysterious as the alignment of the stars. Within a century, we had discovered the existence of viruses, categorized them according to their structure and effects, tracked how they traveled and mutated, and even taken their photographs. We exhumed the 1918 influenza virus in the Arctic and pieced it back together in the lab. We decoded its genetic makeup and, with some controversy, resurrected it. But the medical revolution that began in the mid-1800s and crested with the creation of antibiotics and vaccines will not be complete until we eradicate the flu.

Equally impressive is the way we can now *respond* to the virus. The most important new tools in our war chest have nothing to do with the virus itself. Those tools include antibiotics, which allow us to treat the complications that may follow influenza, as well as intensive care units, respirators ready to take over for beleaguered lungs, and specialists standing by who know the ins and outs of emergency care and infectious diseases. The 1918 virus decimated cities and brought economies to their knees. Then, there was no effective treatment; all that could be done was to offer words of comfort while waiting for the patient to recover or die. Quackery proliferated, and even mainstream therapies like bloodletting were more likely to kill than to cure. That just isn't true today.

But we still don't have a drug that reliably attacks the influenza virus itself. What we have instead are antiviral medications whose benefits are at best controversial, and at worst nonexistent. We urgently need to develop a safe and effective drug that can destroy the virus. We have been working on this goal for decades, but it remains beyond our reach. We can respond, but we still do not have *the response*, the one drug we really need.

In 1918 we reacted. We did not prepare. Today we are much better at preparing for disasters. Every state has a pandemic influenza plan. These plans address everything from obtaining vaccines and coordinating hospitals to setting up ancillary treatment sites in school gymnasia and nursing homes. On a federal level, the Strategic National Stockpile keeps millions of doses of flu vaccines and antiviral medications. The Department of Health and Human Services updated its own Pandemic Influenza Plan in 2017, and it's more than fifty pages long. "Pandemic influenza is not a theoretical threat; rather, it is a recurring threat," the plan's foreword says. "Even so, we don't know when the next pandemic will occur, or how severe it will be."

One of the key components of our preparation is the yearly influenza vaccine, but it is rarely more than 50 percent effective. Although there is agreement that high-risk groups should be vaccinated, we still don't have good enough evidence to drive other policy decisions, like whether healthy adults should be routinely vaccinated. Getting the evidence we need is expensive, but it's a fraction of the cost of stockpiling vaccines and drugs of dubious value.

Our current preparation plan assumes that there will be another pandemic on the scale of 1918, and plenty of experts fear that more than they do most other potential health crises. So why hasn't there been a repeat of 1918 over the last hundred years? Are we allowing the trauma of the past, and anxiety over the future, to make us afraid of a scenario that is highly unlikely, especially given the strides of modern medicine?

There have been two approaches to this question: the pessimistic and the optimistic. For the pessimist, the next flu epidemic looks deadly and inevitable. The news media is full of pessimists because pessimism makes headlines. From printed magazines to cable news to almost every book about influenza that I have read, it seems alarmingly clear that a pandemic is simply unavoidable. Here are the chief reasons the pessimists could be right:

Experts are the ones issuing the warnings. They know influenza and have dedicated their careers to studying the virus. We should take them seriously because they take influenza seriously.

The 1918 pandemic and the outbreaks of 1957 and 1968 confirm that waves of deadly influenza are not theoretical occurrences. It's entirely reasonable to assume that real past outbreaks will translate into real future events.

There have been other recent epidemics—like SARS, Ebola, and Zika—that have given us a sense of what could happen in the future.

These diseases had no respect for borders. Neither does influenza. Now consider the growth in international travel since 1918, when the only way to get from the U.S. to England was a five-day ocean voyage. Today, the same trip takes six hours by plane. We crisscross the globe at incredible speed. And so do the viruses we are carrying.

In spite of all we know about influenza, there is so much that has yet to be figured out. For example, we don't know why the 1918 virus had a predilection for young adults, or why it is a mild disease in some of us but deadly in others. Without understanding these features of the virus, we cannot fully prepare for it.

The domestic poultry population has rapidly increased. Although avian influenza is much more common in wild ducks and geese than it is in domestic poultry, the sheer number of birds that we raise and consume makes transmission more likely. In 2005, the world produced more than 81 million metric tons of poultry meat, the majority of which came from China and Thailand. These are the same countries in which avian flu jumped into the human population. Now add this factor to our rates of international travel and the combination could be deadly.

Influenza loves a crowd. In 1918 the virus quickly infected family members who shared the same cramped, overcrowded homes and apartments. Today, crowded housing is still a fact of life for many in Africa, Asia, and Latin America. And the United States is not immune; there are about 3 million people living in overcrowded conditions. In New York, almost 9 percent of households—more than 280,000 homes—are overcrowded. Even if you live in a spacious house or apartment in the U.S., you might still have to travel to work or school on a crowded subway car or bus that you share with dozens of people each morning.

These factors suggest that a pandemic is inevitable. But before we reach that conclusion, let's give the optimists a chance. There are several solid reasons to believe that the devastation of 1918 was a singular event that will never be exactly repeated. Outbreaks of severe or pandemic influenza are becoming less deadly over time and, as the 2009 swine flu outbreak showed us, their severity can be overestimated. There was, and continues to be, a lot more hype than crisis.

People die from influenza each year, but the number of *excess* deaths—those due to a particularly nasty strain of influenza—has not increased. This may have less to do with our own interventions and more to do with the evolutionary pressures on the virus itself. Viruses love to spread. A virus that is very potent quickly kills its host or sends him to bed; it is then less likely to be transmitted. From the point of view of the influenza virus, the best strategy to reproduce and spread is for it to cause a less severe illness. The newly infected will continue to mingle with the healthy, and this maximizes the chances for the virus to be coughed and sneezed into new hosts. In this respect evolution is on our side. A virus maximizes its chance for reproduction when it is mild. And a mild virus does not a pandemic make.

Another reason to consider the 1918 pandemic a rare occurrence is that certain conditions had to be perfectly aligned for the virus to turn deadly. It had to jump from an avian to a porcine host and from there into humans. This required a specific set of gene swaps and mutations. Had these not been perfect, the virus would not have been nearly as lethal. The conditions also had to be ideal for its transmission. The crowded army camps and military ships in World War I provided those conditions, as did the factories and tenement buildings where the masses worked. The bacterial infections that killed most people would be disarmed by antibiotics today.

Weighing up the evidence leaves me unsure which side to join. Am I a pessimist? An optimist? Each has a compelling case. Each seems entirely plausible to me. Every year in which pandemic influenza fails to appear is another reason for continued optimism—unless you are a pessimist, in which case you'd consider us to be living on borrowed time.

We're certainly more likely to hear pessimistic news. Pessimists are louder. Public health officials make recommendations based on worst-case scenarios. The never-ending news cycle on the internet and cable TV always uses scare tactics to attract our attention. Fearmongering creates anxiety. Don't look for a good prognosis about flu from your news sources, and don't be surprised if we have more influenza seasons like that of 2009, in which the severity and spread of the disease is vastly exaggerated.

The optimists also have a significant problem. Americans, in particular, are a young and optimistic people. They love books on happiness. They love to look away from problems and the negativity of the past. But diseases have histories, and if we don't study the past we can easily find ourselves in danger. It's true that 1918 had the conditions of a perfect storm and that some of those conditions have radically changed since then, but it's equally true that there may be other perfect storms created by a new and unforeseen set of circumstances.

Pessimists complain about the past. Optimists expect a different future. Realists live in the present. They look at the facts and make corrections along the way. When it comes to influenza, I fall in the realist camp. I believe we can look to our encounter with pandemic influenza, use what we now know, and act to prevent it from occurring in the future.

To achieve this, we need to consider one more critical question. It isn't about medicine, science, or policy. Instead it's about our collective memory. Why haven't we done more to acknowledge the history of the flu? While pessimists may be stuck in the past, optimists tend to ignore it. Realists use an awareness of the past to inform the present and future.

Immersed in this research for years, I have come to a realist's conclusion: we have not done enough to place the 1918 pandemic in our collective consciousness. Marking its centennial is a step in the right direction, but it's a very small step. What changes our vigilance about a disease is our capacity as a society to understand its impact, what it has done in the past and how it might affect us in the present. It is certainly the case that research funding helps change outcomes, but what is most important in battling a disease is how widely it is talked about and understood outside of university laboratories and academic seminars.

We memorialize wars, but there are other devastating events that should be part of our collective memory. I'd like to see a memorial in our nation's capital to the flu pandemic of 1918, to honor our losses, to reflect on how far we have come, and to remind us how much more there is yet to do. It's been a century of catastrophe and natural disasters, of world wars and disease and strife. But it's also been a century of mass expansion, of inclusion, of global reach, of technological breakthroughs and medical victories. The flu pandemic tells both these stories. The body in peril and the brain in its element. Humans defeated and humans triumphant. Perhaps by the time a memorial to the 1918 pandemic is built, we will also be celebrating a cure.

ACKNOWLEDGMENTS

Erica Brown, my wife and my best editor. Every intuition you had about this book was spot-on. Without your support in the dark times, it would never have seen the light.

My children, Tali and Yoni, Gavi and Bec, Shai and Alison, and Ayelet. Thank you for giving me the space to write this book, and the love to welcome me back now that it is completed.

Michael Palgon, my agent, for believing in the story and navigating the path from proposal to publication.

Dan Zak, my editor, who gently and expertly guided me and the manuscript through its many draft versions.

Matthew Benjamin, my editor at Simon & Schuster, whose feedback was the map with which I navigated the writing process.

Laura Cherkas, my copy editor and fact-checker extraordinaire. Your eye for detail is simply remarkable.

The following experts patiently allowed me to interview them. Each provided me with all the information I could use and much, much more: Mark Burchess, John Clerici, Peter Doshi, Tom Jefferson, Ali Khan, Gregg Margolis, Brody Mullins, Holt Murray, Forrest Nelson, David Noll, Peter Palese, Andrew Pollard, Autumn Reddinger, Gary Reddinger, Jeff Shaman, Lone Simonsen, Charles Stoecker, Jeff Taubenberger, Don Weiss.

NOTES

PROLOGUE: AUTUMN

1 *Autumn Reddinger was deathly sick:* The details of Autumn's fight with the flu are taken from multiple phone interviews and email correspondence that I had with Autumn, her father, and her physician, Dr. Holt Murray, in December 2017.

6 *kills between 36,000 and 50,000 people each year:* "Estimating Seasonal Influenza-Related Deaths in the United States," Centers for Disease Control and Prevention. Updated January 29, 2018. http://www.cdc.gov/flu/about/disease/us_flu-related_deaths.htm.

6 *more than 2 million people could die:* The estimate of the 1918 U.S. death toll was 675,000 out of a population of about 103 million. Today more than 322 million people live in the United States.

6 *"the worst in nearly a decade":* Donald McNeil, "This Flu Season Is the Worst in Nearly a Decade," *New York Times*, January 27, 2018: A15.

6 *infected 1,400 people between 2012 and 2015:* World Health Organization, "Middle East Respiratory Syndrome Coronavirus (MERS-CoV). Summary of Current Situation, Literature Update and Risk Assessment," July 7, 2015. Available at http://apps.who.int/iris/bitstream/10665/179184/2/WHO_MERS_RA_15.1_eng.pdf.

6 *originated in animal hosts:* SARS seems to have jumped from the Himalayan palm civet population. It's a real animal, and apparently is eaten in China. Here's some more free doctor advice: when next in China, avoid the civet dishes. See W. Li et al., "Animal Origins of the Severe Acute Respiratory Syndrome Coronavirus: Insight from ACE2-S-Protein Interactions," *Journal of Virology* 80, no. 9 (2006): 4211–19.

7   *only effective in about one-third:* B. Flannery et al., "Interim Estimates of 2017–18 Seasonal Influenza Vaccine Effectiveness—United States, February 2018," *Morbidity and Mortality Weekly Report* 67 (2018): 180–185.

1. ENEMAS, BLOODLETTING, AND WHISKEY: TREATING THE FLU

9   *The article was published in 1978:* K. Saketkhoo, A. Januszkiewicz, and M. A. Sackner, "Effects of Drinking Hot Water, Cold Water, and Chicken Soup on Nasal Mucus Velocity and Nasal Airflow Resistance," *Chest* 74, no. 4 (1978): 408–10.

10   *chicken soup is a kind of anti-inflammatory:* B. O. Rennard et al., "Chicken Soup Inhibits Neutrophil Chemotaxis In Vitro," *Chest* 118 no. 4 (2000): 1150–57.

10   *YouTube video:* "Chicken Soup for a Cold." Accessed December 10, 2017. https://www.unmc.edu/publicrelations/media/press-kits/chicken-soup/.

10   *four rounds of bloodletting:* D. M. Morens, "Death of a President," *New England Journal of Medicine* 341, no. 24 (1999): 1845–49.

11   *reanimated by a transfusion of lamb's blood:* Mary Thompson, "Death Defied," George Washington's Mount Vernon. Accessed November 11, 2017. http://www.mountvernon.org/george-washington/the-man-the-myth/death-defied-dr-thorntons-radical-idea-of-bringing-george-washington-back-to-life/.

11   *described a disease:* J. A. B. Hammond, W. Rolland, and T. H. G. Shore, "Purulent Bronchitis: A Study of Cases Occurring amongst the British Troops at a Base in France," *Lancet* 190, no. 4898 (1917): 41–45.

12   *they reported that it worked:* C. E. Cooper Cole, "Preliminary Report on Influenza Epidemic at Bramshott in September-October, 1918," *British Medical Journal* 2, no. 3021 (1918): 566–68. "In some cases venesection relieved the toxemia, especially if combined with (1) saline or (2) glucose and saline interstitially, intravenously, or by the rectum."

12   *"I am an advocate":* Heinrich Stern, *Theory and Practice of Bloodletting* (New York: Rebman Company, 1915), iv.

12   *still advocating bloodletting:* W. F. Petersen and S. A. Levinson, "The Therapeutic Effect of Venesection with Reference to Lobar Pneumonia," *JAMA* 78, no. 4 (1922): 257–58. Petersen and Levinson were real bloodletting groupies. "We believe it but simple justice to many able clinicians of an older period to stress the fact that venesection at times induced striking therapeutic benefits, that a definite and logical basis exists for the therapeutic effects so achieved."

12 *went out of fashion in the twentieth century:* But it took a while. Just how it went out of fashion is discussed in G. B. Risse, "The Renaissance of Bloodletting: A Chapter in Modern Therapeutics," *Journal of the History of Medicine and Allied Sciences* 34, no. 1 (1979): 3–22.

13 *a series of bizarre treatments:* A. F. Hopkirk, *Influenza: Its History, Nature, Cause and Treatment* (New York: The Walter Scott Publishing Company, 1914), 155. I am probably being too harsh here, since nearly all physicians would treat illnesses—of practically any variety—in the same way, using laxatives and purges. See David Wootton, *Bad Medicine: Doctors Doing Harm since Hippocrates* (Oxford: Oxford University Press, 2006).

13 *dying of aspirin overdoses:* In February 1917 the aspirin manufacturer Bayer lost its patent on the drug. This allowed other manufacturers to produce it and flood the market, making large doses more available to people desperate for any kind of treatment. In September 1918 the U.S. surgeon general noted that aspirin had been used successfully in foreign countries to relieve a range of symptoms. The very next month there was a peak in the number of deaths from influenza.

13 *"drenched with aspirin":* Richard Collier, *The Plague of the Spanish Lady: The Influenza Pandemic of 1918–1919* (London: Macmillan, 1974), 106.

13 *help explain the deaths:* See K. M. Starko, "Salicylates and Pandemic Influenza Mortality, 1918–1919 Pharmacology, Pathology, and Historic Evidence," *Clinical Infectious Diseases* 49, no. 9 (2009): 1405–10. See also John M. Barry, *The Great Influenza: The Epic Story of the Deadliest Plague in History* (New York: Penguin, 2005), 353, 358.

14 *"a teaspoonful of Friar's balsam":* Hopkirk, *Influenza*, 159.

14 *also prescribed quinine:* Ibid., 156.

15 *known as Jesuits' powder:* D. C. Smith, "Quinine and Fever: The Development of the Effective Dosage," *Journal of the History of Medicine and Allied Sciences* 31, no. 3 (1976): 343–67.

15 *first-line treatment for malaria:* World Health Organization, "Guidelines for the Treatment of Malaria," Geneva, World Health Organization, 2015.

15 *why not use it to treat all fevers?:* Smith, "Quinine and Fever."

15 *used in England:* Cooper Cole, "Preliminary Report on Influenza Epidemic at Bramshott."

15 *the United States:* H. A. Klein, "The Treatment of 'Spanish Influenza,'" *JAMA* 71, no. 18 (1918): 1510.

15 *the European continent:* ". . . il etait logique d'avoir recours aux injections pour traiter cette infection comme on le fait pour le paludisme." See F. Fabier, "Traitement de la Grippe par les Injections de Quinine," *Journal de Méde-*

cine et de Chirurgie Pratiques 90 (1919): 783–84, and more generally M. L. Hildreth, "The Influenza Epidemic of 1918–1919 in France: Contemporary Concepts of Aetiology, Therapy, and Prevention," *Social History of Medicine* 4, no. 2 (1991): 277–94.

15 *"ringing in the head"*: See, for example, the *Muskogee Times-Democrat*, December 1, 1919, 6.

15 *in high doses:* M. E. Boland, S. M. Roper, and J. A. Henry, "Complications of Quinine Poisoning," *Lancet* 1, no. 8425 (1985): 384–85.

15 *small doses of dry champagne:* Hopkirk, *Influenza*, 163, 180.

15 *no finer pick-me-up:* Ibid., 167.

16 *could not hide his contempt:* "Influenza: Its History, Nature, Cause and Treatment," book review, *JAMA* 63, no. 3 (1914): 267. I still cannot be sure for whom the reviewer had more contempt, the British or Dr. Hopkirk. I love that word "nostrum." It means an ineffective medicine, prepared by an unqualified person.

16 *published seventy years later:* R. J. Sherertz and H. J. Sherertz, "Influenza in the Preantibiotic Era," *Infectious Diseases in Clinical Practice* 14, no. 3 (2006): 127.

17 *or dicing onions to sterilize a room:* Roger Welsch, *A Treasury of Nebraska Pioneer Folklore* (Lincoln: University of Nebraska Press, 1967), 370.

17 *claimed that he had treated 225 patients:* "Influenza Discussions," *American Journal of Public Health* 9, no. 2 (1919): 136.

17 *Captain A. Gregor, set off to investigate this claim:* A. Gregor, "A Note on the Epidemiology of Influenza among Workers," *British Medical Journal* 1, no. 3035 (1919): 242–43.

18 *poison gas workers:* F. Shufflebotham, "Influenza among Poison Gas Workers," *British Medical Journal* 1, no. 3042 (1919): 478–79. For some reason this immunity did not extend to those who worked with phosgene gas, which had been used to horrible effect in the trenches of World War I.

18 *viruses floating around those gasworks:* E. W. Rice et al., "Chlorine Inactivation of Highly Pathogenic Avian Influenza Virus (H5N1)," *Emerging Infectious Diseases* 13, no. 10 (2007): 1568–70.

18 *a brilliant physician:* "James B. Herrick (1861–1954)," *JAMA* 16, no. 186 (1963): 722–23.

18 *laid the foundations of modern cardiology:* See C. S. Roberts, "Herrick and Heart Disease," in H. Kenneth Walker, W. Dallas Hall, and J. Willis Hurst, eds., *Clinical Methods: The History, Physical, and Laboratory Examinations*, 3rd ed. (Atlanta: Butterworth Publishers, 1990). Another notable review of Herrick's work is R. S. Ross, "A Parlous State of Storm and Stress. The Life and Times of James B. Herrick," *Circulation* 67, no. 5 (1983): 955–59.

19 *his plea was simple:* James B. Herrick, "Treatment of Influenza by Means Other Than Vaccines and Serums," *JAMA* 73, no. 7 (1919): 482–87.

19 *"some one has blundered in reaching conclusions":* All the quotations are from Herrick, 483.

19 *squarely in the conservative mainstream:* "Proceedings of the Forty-Sixth Annual Meeting of the American Public Health Association," *American Journal of Public Health* 9, no. 2 (1919): 130–42.

19 *addressed the use of laxatives:* The quote is from Herrick, 483. This focus on the bowel movements of patients with influenza was incredibly widespread and part of the medical establishment's ingrained wisdom. Here's part of a physician's letter published in the *Journal of the American Medical Association* in November 1918: "I cannot too strongly call attention to the importance of keeping the bowels open by mild cathartics. Often a lingering temperature disappeared after a brisk flushing of the bowels, and this method greatly aided in shortening the course of the disease." From Klein, "The Treatment of 'Spanish Influenza,'" 1510.

20 *wrote that from the large number of agents:* Cooper Cole, "Preliminary Report on Influenza Epidemic at Bramshott."

21 *what happens more than 31 million times:* N. A. Molinari et al., "The Annual Impact of Seasonal Influenza in the U.S.: Measuring Disease Burden and Costs," *Vaccine* 25, no. 27 (2007): 5086–96.

22 *the early-morning hours are usually quiet:* These observations are based on twenty-five years of my own experience of working in a couple of dozen emergency departments in the U.S. and abroad. Fortunately, my experiences seem to match perfectly with the published data. My former colleague Melissa McCarthy studied emergency department arrivals at a large urban teaching hospital over one year. She found that Mondays and Fridays were the busiest days, and that early-morning hours were the slowest. See M. L. McCarthy et al., "The Challenge of Predicting Demand for Emergency Department Services," *Academic Emergency Medicine* 15, no. 4 (2008): 337–46. See also S. J. Welch, S. S. Jones, and T. Allen, "Mapping the 24-Hour Emergency Department Cycle to Improve Patient Flow," *Joint Commission Journal on Quality and Patient Safety* 33, no. 5 (2007): 247–55. These patterns are found in emergency departments all over the world. See, for example, Y. Tiwari, S. Goel, and A. Singh, "Arrival Time Pattern and Waiting Time Distribution of Patients in the Emergency Outpatient Department of a Tertiary Level Health Care Institution of North India," *Journal of Emergencies, Trauma, and Shock* 7, no. 3 (2014): 160–65.

22 *medical teams are likely to be at their slowest at the end of their shifts:* Most

emergency departments have three main shifts: 7 a.m. to 3 p.m., 3 p.m. to 11 p.m., and 11 p.m. to 7 a.m. In addition, there are numerous combinations of additional overlapping shifts that vary depending on the peak patient arrival times for a particular ER.

23 *jury awarded the estate:* A. Elliott-Engel, "Jury Awards $1.87 Million in Carbon Monoxide Poisoning Case," *Legal Intelligencer,* June 1, 2011.

24 *so we treat them:* See, for example, M. Glatstein and D. Scolnik, "Fever: To Treat or Not to Treat?," *World Journal of Pediatrics* 4, no. 4 (2008): 245–47. A dated but useful review of the subject is Matthew J. Kluger, *Fever: Its Biology, Evolution, and Function* (Princeton, NJ: Princeton University Press, 1979).

24 *improves the efficacy of another group of blood cells:* A review of fever and the thermal regulation of immunity concluded that "febrile temperatures serve as a systemic alert system that broadly promotes immune surveillance during challenge by invading pathogens." See S. S. Evans, E. A. Repasky, and D. T. Fisher, "Fever and the Thermal Regulation of Immunity: The Immune System Feels the Heat," *National Review of Immunology* 15, no. 6 (2015): 335–49.

24 *49,000 people die from the flu each year:* These are the CDC estimates. See "Estimating Seasonal Influenza-Related Deaths in the United States."

24 *plug the McMaster estimates into these flu numbers:* D. J. Earn, P. W. Andrews, and B. M. Bolker, "Population-Level Effects of Suppressing Fever," *Proceedings of the Royal Society B: Biological Sciences* 281, no. 1778 (2014): 20132570.

25 *they are able to walk out of the ER:* Intravenous fluids are a simple intervention, with manufacturers charging about one dollar for a bag. But that doesn't prevent hospitals from charging the patient an incredible markup. An investigation in the *New York Times* revealed that some people were charged as much as $787 for "IV therapy." In one case, a patient was charged $91 for IV fluid that cost the hospital $0.86. And you thought hotel minibars were a rip-off. See Nina Bernstein, "How to Charge $546 for Six Liters of Saltwater," *New York Times,* August 27, 2013.

27 *get a completely useless antibiotic:* C. G. Grijalva, J. P. Nuorti, and M. R. Griffin, "Antibiotic Prescription Rates for Acute Respiratory Tract Infections in U.S. Ambulatory Settings," *JAMA* 302, no. 7 (2009): 758–66.

2. THE JOLLY RANT: A HISTORY OF THE VIRUS

29 *They exist on the edge of life:* E. Rybicki, "The Classification of Organisms at the Edge of Life, or Problems with Virus Systematics," *South African Journal of Sciences* 86 (1990): 182–98.

29 *incorporated into our own genetic code:* M. Emerman and H. S. Malik, "Paleovirology—Modern Consequences of Ancient Viruses," *PLoS Biology* 8, no. 2 (2010): e1000301.

30 *a Latin word:* Sally Smith Hughes, *The Virus: A History of the Concept* (London: Heinemann Educational Books, Science History Publications, 1977), 109–14.

30 *Edward Jenner:* ". . . what renders the Cow-pox virus so extremely singular, is, that the person who has been thus affected is for ever after secure from the infection of the Small Pox." Edward Jenner, *An Inquiry into the Causes and Effects of the Variolae Vaccinae, a Disease Discovered in Some of the Western Counties of England, Particularly Gloucestershire, and Known by the Name the Cow Pox* (London: Sampson Low, 1798), 6. We will return to Jenner in chapter 9.

30 *"le virus rabique":* Hughes, *The Virus,* 112.

30 *They contain a core of genetic material:* Ibid., 114.

30 *Thucydides described:* As cited in A. D. Langmuir et al., "The Thucydides Syndrome. A New Hypothesis for the Cause of the Plague of Athens," *New England Journal of Medicine* 313, no. 16 (1985): 1027–30.

31 *wrote Hippocrates:* Francis Adams, *The Genuine Works of Hippocrates* (New York: William Wood, 1886), 298.

31 *a flu epidemic broke out:* Charles Creighton, *A History of Epidemics in Britain,* 2nd ed., vol. 2 (New York: Barnes & Noble, 1965), 328. Creighton records earlier epidemics of coughing, like that which occurred in April 1658. However, influenza is usually a disease of the winter, and it is therefore very unlikely to have been the cause of that April epidemic.

32 *impossible to hear the sermon:* Ibid., 328.

32 *Thomas Sydenham suggested:* Ibid., 329.

32 *Bloodletting and laxatives:* The epidemic in the winter of 1729 seems to have been especially brutal, and affected Britain, Ireland, and, later, Italy. See ibid., 343. Not all of the epidemics mentioned by Creighton were from influenza. For example, there was an epidemic in April 1743 in which "the skin was very frequently inflamed when the fever ran high; and it afterwards peeled off in most parts of the body." This is not a description of what we now call viral influenza, and was far more likely to have been what was once commonly called scarlet fever, caused by a streptococcal bacterial infection.

32 *a headline in the* New York Times: Lawrence Altman, "Is This a Pandemic? Define 'Pandemic,'" *New York Times,* June 8, 2009, D1. See also D. M. Morens, G. K. Folkers, and A. S. Fauci, "What Is a Pandemic?," *Journal of Infectious Disease* 200, no. 7 (2009): 1018–21.

32 *The most useful definition we have:* K. D. Patterson, *Pandemic Influenza, 1700–1900* (Totowa, NJ: Rowman and Littlefield, 1986), 5.

32 *Some of these outbreaks:* Ibid., 83.

33 *Henry Parsons reported on it:* Henry Franklin Parsons, *Report on the Influenza Epidemic of 1889–90* (London, Eyre and Spottiswoode, 1891).

33 *spread to the United States:* Ibid., 24–27.

33 *even a rumor:* Ibid., 107. Elsewhere (p. 102) Parsons cites a French professor who believed "that Influenza is a growth of Russian soil, and when not a raging malady is a smoldering one." Parsons, however, was skeptical of this claim, noting that the conditions that existed in Russia were similar to those in other parts of Europe: "If such conditions may prove a breeding ground for Influenza in Russia, it may be asked why not elsewhere?" It is ironic that I am writing these words during an FBI investigation into Russia's role in influencing the 2016 U.S. presidential election. Is there nothing for which the Russians don't get blamed?

33 *caused by the conjunction of Jupiter and Saturn:* Creighton, *A History of Epidemics in Britain,* 397–409.

33 *three possible origins:* Parsons, *Report on the Influenza Epidemic of 1889–90,* 70.

34 *"non-living particulate material":* Ibid., 82.

34 *rates were higher among the clerks:* Ibid., 73.

34 *Parsons was convinced:* Ibid., 102.

34 *animals were somehow responsible:* Ibid., 106.

35 *scientists identified the specific germs:* Hughes, *The Virus,* 6–8.

35 *claimed to have discovered the bacterium:* J. K. Taubenberger, J. V. Hultin, and D. M. Morens, "Discovery and Characterization of the 1918 Pandemic Influenza Virus in Historical Context," *Antiviral Therapy* 12, no. 4, part B (2007): 581–91.

36 *"an authoritative road sign":* Alfred W. Crosby, *America's Forgotten Pandemic: The Influenza of 1918,* 2nd ed. (Cambridge: Cambridge University Press, 2003), 269.

37 *simple virus to depict:* J. K. Taubenberger, A. H. Reid, and T. G. Fanning, "Capturing a Killer Flu Virus," *Scientific American* 292, no. 1 (January 2005): 62–71.

41 *"A disease of undetermined nature":* "Undetermined Disease—Valencia," *Public Health Reports* 33, no. 26 (1918): 1087.

41 *buried among news of the fighting:* "Spanish Influenza Is Raging in the German Army," *New York Times,* June 27, 1918.

41 *the German Kaiser himself would catch it:* In a one-sentence dispatch, the paper reported that "the Kaiser and Kaiserin are suffering from the Span-

ish influenza in a mild form." "Kaiser Has Influenza," *New York Times*, July 19, 1918.

3. "SOMETHING FIERCE": THE SPANISH FLU OF 1918

43 *a report to health officials:* "Influenza. Kansas—Haskell," *Public Health Reports* 33, no. 14 (1918): 502.

43 *first recorded instance:* J. M. Barry, "The Site of Origin of the 1918 Influenza Pandemic and Its Public Health Implications," *Journal of Translational Medicine* 2, no. 1 (2004): 1–4.

44 *the U.S. Army's Camp Funston:* Ibid.

44 *the virus expanded outward in waves:* Ibid. For a general description of the outbreak at Camp Funston, see E. L. Opie et al., "Pneumonia at Camp Funston," *JAMA* 72, no. 2 (1919): 108–113.

44 *first at Brest:* Barry, "The Site of Origin of the 1918 Influenza Pandemic"; F. M. Burnet and E. Clark, *Influenza: A Survey of the Last 50 Years in the Light of Modern Work on the Virus of Epidemic Influenza*, monograph from the Walter and Eliza Hall Institute for Research in Pathology and Medicine (Melbourne: Macmillan and Company, 1942), 70–71.

44 *hit between September and November 1918:* J. S. Oxford, "The So-Called Great Spanish Influenza Pandemic of 1918 May Have Originated in France in 1916," *Philosophical Transactions of the Royal Society of London, Series B: Biological Sciences* 356, no. 1416 (2001): 1857–59.

44 *photographs of French soldiers:* Ibid. Oxford specifically addressed the question of whether the virus originated in China. He thought that this possibility, while it could not be excluded, was "unlikely" (p. 1859).

44 *"a curious epidemic resembling influenza is sweeping over North China":* "Queer epidemic sweeps North China," *New York Times*, June 1 1918, 1.

45 *predated the general outbreaks:* Christopher Langford, "Did the 1918–19 Influenza Pandemic Originate in China?," *Population and Development Review* 31, no. 3 (2005): 473–505; K. F. Shortridge, "The 1918 'Spanish' Flu: Pearls from Swine?," *Nature Medicine* 5, no. 4 (1999): 384–85.

45 *a precursor to the 1918 flu:* Shortridge raises the possibility that "an II1-like virus may have existed in humans, at least in southern China, for some 50 years before the earliest epidemiological evidence of its presence there."

45 *140,000 Chinese laborers were recruited:* Langford, "Did the 1918–19 Influenza Pandemic Originate in China?"

45 *the British Army's camp at Etaples:* Shortridge cites Lynn MacDonald as the source for the claim that there were Chinese workers near Etaples. See

NOTES

Lynn MacDonald, *Somme* (London: Macmillan, 1984), 189–93. An exhaustively detailed analysis of the Chinese origin theory is found in Langford. He concludes that "the finding that influenza was widespread in China in 1918–19, but at the same time, in many parts of the country, less lethal than elsewhere in the world, despite the generally poor levels of health there at the time, suggests very strongly, Oxford and others notwithstanding, that the 1918–19 influenza virus originated in China" ("The 1918 'Spanish' Flu," 494). Shortridge is certain of this theory: "I believe that southern China was the source of the virus, in keeping with the hypothesis that the region is a hypothetical influenza epicenter for the emergence of pandemic influenza viruses, and it spread with the economy-driven movement of people out of Guangdong Province" ("Did the 1918–19 Influenza Pandemic Originate in China?," 385).

45 *did not first erupt in Spain:* But the name has stuck. In his excellent history of the 1918 pandemic, Alfred Crosby refers to the influenza as "Spanish" at least forty-seven times, even though the second edition of the book was published in 2003. See Crosby, *America's Forgotten Pandemic*. Richard Collier's 1974 book on the epidemic left no ambiguity; it was titled *The Plague of the Spanish Lady* (London: Macmillan, 1974).

46 *attacked in two waves:* Details come from Carol Byerly, *Fever of War: The Influenza Epidemic in the U.S. Army during World War I* (New York: New York University Press, 2005) and Barry, *The Great Influenza*. Some academics have counted four or more waves.

47 *propagated the bacteria:* G. D. Shanks et al., "Variable Mortality from the 1918–1919 Influenza Pandemic during Military Training," *Military Medicine* 181, no. 8 (2016): 878–82.

47 *around 60 died:* There is some disagreement between the numbers in Shanks, and the numbers cited by Barry in *The Great Influenza* and by Crosby.

47 *His soldiers weren't able to march:* Byerly, *Fever of War*, 72.

47 *31,000 cases of influenza among British troops:* Crosby, *America's Forgotten Pandemic*, 26.

47 *troops were unable to report for combat:* Barry, *The Great Influenza*, 174.

48 *3 percent had died:* See Crosby, *America's Forgotten Pandemic*, 38.

49 *the* British Medical Journal *reported that influenza was no longer a threat:* "The many complications and sequelae by which in times gone by epidemic influenza made itself remembered seem to be happily rare. This circumstance has been held to show that we are not now dealing with an epidemic or a pandemic of influenza at all." The anonymous author wistfully yearned for earlier days. "How much better it would be for all of us,

and how fatal to the spread of influenza," he wrote, "if we could all go back to our childhood, and learn once more with the thoroughness engendered by the fear of, say, an instant smacking never to cough or sneeze without first covering both mouth and nose with a handkerchief! But this is to ask too much." See "The Influenza Pandemic," *British Medical Journal* 2 (1918): 39.

49 *earliest reports of the second wave:* Barry, *The Great Influenza*, 186.

50 *a letter dated September 29, 1918:* The letter was found along with other medical papers in a trunk in Detroit, and eventually wound up in the Department of Epidemiology at the University of Michigan. See N. R. Grist, "Pandemic Influenza 1918," *British Medical Journal* 2, no. 6205 (1979): 1632–33.

51 *"demonstrated the inferiority of human interventions":* Victor C. Vaughan, *Doctor's Memories* (New York: Bobbs-Merrill Company, 1926), 384.

51 *sickened 14,000 and left 750 dead:* Byerly, *Fever of War*, 75–76.

51 *four times its maximum capacity:* Ibid., 84.

51 *life would slowly return to normal:* Ibid., 76.

51 *the death toll in the U.S. stood at 675,000:* N. P. Johnson and J. Mueller, "Updating the Accounts: Global Mortality of the 1918–1920 'Spanish' Influenza Pandemic," *Bulletin of the History of Medicine* 76 (2002): 105–15.

52 *the leap from army camps to civilian neighborhoods:* Information is taken from Barry, *The Great Influenza*, 200–227.

52 *German submarines loaded with germs:* Cited in Gina Kolata, *Flu: The Story of the Great Influenza Pandemic of 1918 and the Search for the Virus That Caused It* (New York: Touchstone, 2005), 3.

52 *masquerading under a new name:* Ibid.

52 *an article in the* Philadelphia Inquirer: "Big Pageant to Launch Philadelphia's Fourth Loan Drive," *Philadelphia Inquirer*, September 28, 1918, 3.

53 *"a tremendously impressive pageant":* "Representatives of a Great Nation Embattled Take Part in Tremendously Impressive Pageant," *Philadelphia Inquirer*, September 29, 1918, 15.

54 *"compelled by the epidemic":* Crosby, *America's Forgotten Pandemic*, 75.

54 *Some funeral homes:* Ibid., 83.

55 *More than 1,000 people died:* Ibid., 99.

55 *imposing a quarantine:* Ibid., 239–57.

55 *a fight against germs and Germans:* Byerly, *Fever of War*, 97.

56 *688 were hospitalized and 49 died:* Barry, *The Great Influenza*, 173.

56 *4,000 Parisians died:* Ibid., 362.

56 *"It was a grievous business":* Cited in Byerly, *Fever of War*, 73.

56 *agreed to limit any discussion of the flu:* Mark Honigsbaum, "Regulating the

1918–19 Pandemic: Flu, Stoicism and the Northcliffe Press," *Medical History* 57, no. 2 (2013): 165–85.

56   *embodied in a letter:* J. McOscar, "Influenza in the Lay Press," *British Medical Journal* 2, no. 3019 (1918): 534.

57   *a catastrophic outbreak among British and French troops:* Influenza Committee of the Advisory Board to the D.G.M.S. France, "The Influenza Epidemic in the British Armies in France, 1918," *British Medical Journal* 2, no. 3019 (1918): 505–9.

57   *His advice was limited:* Juliet Nicolson, *The Great Silence, 1918–1920: Living in the Shadow of the Great War* (London: Grove Press, 2010), 93.

57   *"more stoically accepted":* Anon., "6,000,000 Deaths. Influenza World Toll," *Times* (London), December 18, 1918, 5.

57   *"cheerfully anticipating":* Anon., "The Spanish Influenza. A Sufferer's Symptoms," *Times* (London), June 25, 1918, 9.

57   *Over 225,000 died:* Ben Johnson, "The Spanish Flu Epidemic of 1918," Historic UK. Accessed April 25, 2018. https://www.historic-uk.com/HistoryUK/HistoryofBritain/The-Spanish-Flu-pandemic-of-1918/.

57   *20 million Indians died:* Barry, *The Great Influenza*, 364.

58   *The first explanation:* Taubenberger, Reid, and Fanning, "Capturing a Killer Flu Virus."

59   *"cytokine storm":* D. M. Morens and A. S. Fauci, "The 1918 Influenza Pandemic: Insights for the 21st Century," *Journal of Infectious Disease* 2007, no. 195 (2007): 1019–28.

59   *the biggest unsolved mystery of the pandemic:* Ibid.

60   *the death rate was twenty-five times greater:* C. J. Murray et al., "Estimation of Potential Global Pandemic Influenza Mortality on the Basis of Vital Registry Data from the 1918–20 Pandemic: A Quantitative Analysis," *Lancet* 368, no. 9554 (2006): 2211–18.

60   *that the average life expectancy in 1918:* D. W. Smith and B. S. Bradshaw, "Variation in Life Expectancy during the Twentieth Century in the United States," *Demography* 43, no. 4 (2006): 647–57. This twelve-year drop should be contrasted with the ongoing opioid epidemic, which also caused a drop in the average life expectancy, but this time by one-tenth of a year. See K. D. Kochanek et al., "Mortality in the United States, 2016," *NCHS Data Brief*, no. 293 (2017): 1–8.

60   *It makes for terrifying reading:* G. M. Price, "Influenza—Destroyer and Teacher," *Survey* 41, no. 12 (1918): 367–69.

61   *"each diseased person in a diver's suit":* Ibid., 367.

61   *Chicago's health commissioner:* Ibid., 368.

62 *where they could be taught how to stay healthy:* Natalie S. Robins, *Copeland's Cure: Homeopathy and the War between Conventional and Alternative Medicine,* 1st ed. (New York: Knopf, 2005), 151.

63 *"never again to prate":* As cited in Price, "Influenza—Destroyer and Teacher," 367.

63 *T. Yamanouchi:* T. Yamanouchi, K. Sakakami, and S. Iwashima, "The Infecting Agent in Influenza: An Experimental Research," *Lancet* 193, no. 4997 (1919): 971.

64 *when given to live rabbits:* P. Olitsky and F. Gates, "Experimental Study of the Nasopharyngeal Secretions from Influenza Patients," *JAMA* 74, no. 22 (1920): 1497–99.

64 *Soon there were reports:* Taubenberger, Hultin, and Morens, "Discovery and Characterization of the 1918 Pandemic Influenza Virus in Historical Context." These included a lot of animal diseases: foot-and-mouth disease in cattle, bovine pneumonia, rabbit myxomatosis, and African horse sickness.

64 *eliminated bacteria as a suspect:* M. C. Winternitz, I. M. Wason, and F. P. McNamara, *The Pathology of Influenza* (New Haven, CT: Yale University Press, 1920), 55.

64 *British scientists concluded:* W. Smith, C. H. Andrewes, and P. P. Laidlaw, "A Virus Obtained from Influenza Patients," *Lancet* 2, no. 5723 (1933): 66–68. The existence of the virus was, at this stage, an assumption. It really meant an infectious particle small enough to pass through the filters that strain out bacteria.

64 *the discovery that the influenza virus could be cultivated:* Hughes, *The Virus,* 93.

4. "AM I GONNA DIE?": ROUND TWO, AND THREE, AND FOUR . . .

67 *700,000 refugees:* "Hong Kong Battling Influenza Epidemic," *New York Times,* April 17, 1957, 3.

67 *this outbreak was caused by a different strain:* E. D. Kilbourne, "Influenza Pandemics of the 20th Century," *Emerging Infectious Diseases* 12, no. 1 (2006): 9–14.

68 *more than 60 percent became clinically ill:* D. A. Henderson et al., "Public Health and Medical Responses to the 1957–58 Influenza Pandemic," *Biosecurity and Bioterrorism* 7, no. 3 (2009): 265–73.

69 *Hilleman went to work:* Lawrence Altman, "Maurice Hilleman, Master in Creating Vaccines, Dies at 85," *New York Times,* April 12, 2005.

69 *He recalled later:* "1957 Asian Flu Pandemic," video interview with Maurice Hilleman, The History of Vaccines. Accessed April 25, 2018. https://www.historyofvaccines.org/content/1957-asian-flu-pandemic.

69 *It quickly spread to Southeast Asia:* W. C. Cockburn et al., "Origin and progress of the 1968–69 Hong Kong influenza epidemic," *Bulletin of the World Health Organization* 41 (1969): 343–48.

70 *provided some immunity against the Hong Kong flu:* Kilbourne, "Influenza Pandemics of the 20th Century."

70 *was still less deadly than the Asian flu:* P. R. Saunders-Hastings and D. Krewski, "Reviewing the History of Pandemic Influenza: Understanding Patterns of Emergence and Transmission," *Pathogens* 5, no. 4 (2016): 1–19.

70 *Center for Disease Control:* The organization was called the Center for Disease Control until 1980, when it was given a plural title, the Centers for Disease Control. In 1992, Congress tweaked its name again. It became the Centers for Disease Control *and Prevention.*

71 *it was of pivotal importance:* Details are from J. C. Gaydos et al., "Swine Influenza A Outbreak, Fort Dix, New Jersey, 1976," *Emerging Infectious Diseases* 12, no. 1 (2006): 23–28, and Gina Kolata, *Flu: The Story of the Great Influenza Pandemic of 1918 and the Search for the Virus That Caused It* (New York: Touchstone, 2005), chapters 5–6.

72 *He also called on the CDC:* E. Kilbourne, "Flu to the Starboard! Man the Harpoons!," *New York Times*, February 13, 1976, 33.

72 *"Better a vaccine without an epidemic":* Kolata, *Flu*, 139.

72 *in a letter to the* New York Times: Hans Neumann, "After the Flu Shots," *New York Times*, September 15, 1976, 44.

73 *Ford said in the White House:* Gerald Ford, XXXVIII President of the United States, "Remarks Announcing the National Swine Flu Immunization Program." Accessed April 25, 2018. http://www.presidency.ucsb.edu/ws/index.php?pid=5752. It is not clear where Ford got the number of 548,000 deaths.

74 *Walter Cronkite appeared on the evening news:* D. J. Sencer and J. D. Millar, "Reflections on the 1976 Swine Flu Vaccination Program," *Emerging Infectious Diseases* 12, no. 1 (2006): 29–33.

74 *to kill the head of the Gambino crime family:* Cited in Kolata, *Flu*, 165.

74 *A "sorry debacle":* Harry Schwartz, "Swine Flu Fiasco," *New York Times*, December 21, 1976, 33.

74 *This was "the swine flu snafu":* Matt Clark, "The Swine Flu Snafu," *Newsweek*, July 12, 1976, 73.

75 *CDC's director, was forced to resign:* Bruce Weber. "David J. Sencer Dies at 86; Led Disease-Control Agency," *New York Times*, May 4, 2011, A27.

75 *and 23 of those people had died:* Douglas Martin, "Edwin Kilbourne, Flu Vaccine Expert, Dies at 90," *New York Times*, February 24, 2011, B14.

75 *on its seasonal influenza website:* "Guillain-Barré Syndrome and Flu Vaccine,"

Centers for Disease Control and Prevention. Accessed May 2, 2018. https://www.cdc.gov/flu/protect/vaccine/guillainbarre.htm. The CDC's wording has evolved. As recently as 2017, the same website stated that in 1976 there had been "a small increased risk of GBS following vaccination with an influenza vaccine made to protect against a swine flu virus." See Internet Archive Wayback Machine, "Guillain-Barré Syndrome and Flu Vaccine." https://web.archive.org/web/20170508051122/https://www.cdc.gov/flu/protect/vaccine/guillainbarre.htm. Accessed April 25, 2018.

75 *Sencer stood by the decision:* Sencer and Millar, "Reflections on the 1976 Swine Flu Vaccination Program."

75 *Kilbourne called for better influenza preparation:* All the quotes in this paragraph are from Kilbourne, "Influenza Pandemics of the 20th Century."

76 *In March 2009:* The timeline and data are based on "The 2009 H1N1 Pandemic: Summary Highlights, April 2009–April 2010," Centers for Disease Control and Prevention. Accessed April 25, 2018. https://www.cdc.gov/h1n1flu/cdcresponse.htm#CDC_Laboratories_Bolster_Nations_Testing.

76 *the strain contained genes from four ancestors:* Khan and William Patrick, *The Next Pandemic: On the Front Lines against Humankind's Gravest Dangers,* 1st ed. (New York: PublicAffairs, 2016), 24.

76 *director-general of the World Health Organization (WHO) declared it a pandemic:* Margaret Chan, "World now at the start of 2009 influenza pandemic," World Health Organization. Accessed April 25, 2018. http://www.who.int/mediacentre/news/statements/2009/h1n1_pandemic_phase6_20090611/en/.

77 *The FDA received 1,371 requests:* A. Sorbello et al., "Emergency Use Authorization for Intravenous Peramivir: Evaluation of Safety in the Treatment of Hospitalized Patients Infected with 2009 H1N1 Influenza A Virus," *Clinical Infectious Diseases* 55, no. 1 (2012): 1–7.

78 *"People need to understand that this vaccine is safe":* Jesse Lee, "The President and First Lady Get Vaccinated," The White House. Accessed April 25, 2019. https://obamawhitehouse.archives.gov/blog/2009/12/21/president-and-first-lady-get-vaccinated.

78 *"If I had the two people that are most important in my life":* Ibid.

78 *an unusually low figure for an influenza outbreak:* Peter Doshi, "The 2009 Influenza Pandemic," *Lancet Infectious Diseases* 13, no. 3 (2013): 193–94. Final estimates of deaths in the 2009 pandemic are from S. S. Shrestha et al., "Estimating the Burden of 2009 Pandemic Influenza A (H1N1) in the United States (April 2009–April 2010)," *Clinical Infectious Diseases* 52, suppl. 1 (2011): s75–s82.

78 *"'Am I gonna die?'":* Kathryn Tolbert, "Local Teens Describe Their Expe-

riences With Swine Flu," The Washington Post.com, August 25 2009. Accessed April 25, 2018. http://www.washingtonpost.com/wp-dyn/content/article/2009/08/24/AR2009082402346.html.

78 *there was no cause for alarm:* Robert Peer and Gardiner Harris, "Obama Seeks to Ease Fears on Swine Flu," *New York Times,* April 27, 2009, A1.

78 *declared the H1N1 outbreak to be a national emergency:* "President Obama Signs Emergency Declaration for H1N1 Flu," The White House. Accessed April 25, 2018. https://obamawhitehouse.archives.gov/blog/2009/10/25/president-obama-signs-emergency-declaration-h1n1-flu.

79 *It now had its own hashtag:* Martin Szomszor, Patty Kostkova, and Ed de Quincey, "#Swineflu: Twitter Predicts Swine Flu Outbreak in 2009," in *Electronic Healthcare* (New York: Springer, 2011): 18–26.

79 *CNN and Fox News were criticized:* John Sutter, "Swine flu creates controversy on Twitter," CNN. Accessed April 25, 2018. http://www.cnn.com/2009/TECH/04/27/swine.flu.twitter/.

79 *Fiona Godlee, editor of the influential* British Medical Journal: Fiona Godlee, "Conflicts of Interest and Pandemic Flu," *British Medical Journal* 340 (2010): c2947.

79 *the WHO's official definition:* Cited in Peter Doshi, "The Elusive Definition of Pandemic Influenza," *Bulletin of the World Health Organizaton* 89 (2011): 532–38.

79 *"It was a mistake, and we apologize":* Elizabeth Cohen, "When a Pandemic Isn't a Pandemic," CNN. Accessed April 25, 2018. http://edition.cnn.com/2009/HEALTH/05/04/swine.flu.pandemic/index.html.

80 *the description of a pandemic did not match its severity:* All the examples in this paragraph are taken from Table 2 of Doshi, "The Elusive Definition of Pandemic Influenza."

80 *John Barry noted:* John Barry, "Lessons from the 1918 Flu," *Time,* October 17, 2005, 96.

5. RESURRECTING THE FLU

82 *pressure to show Congress:* But in the end, the institute was closed. See Christopher Lee, "Pathologists Protest Defense Site's Closure," *Washington Post,* February 4, 2007.

83 *not contain any viral particles:* J. K. Taubenberger et al., "Initial Genetic Characterization of the 1918 'Spanish' Influenza Virus," *Science* 275, no. 5307 (1997): 1793–96.

84 *later became a division of the AFIP:* Byerly, *Fever of War,* 181.

84 *features in common with bird flu:* J. K. Taubenberger, "The Origin and Virulence of the 1918 'Spanish' Influenza Virus," *Proceedings of the American Philosophical Society* 150, no. 1 (2006): 86–112.

85 *"It was a gimmick":* Interview with Jeffrey Taubenberger, June 14, 2016. The story of how Taubenberger came to sequence the 1918 flu virus is recalled slightly differently in Kolata, *Flu*, 193–208.

85 *the sequence of these four letters:* Taubenberger et al., "Initial Genetic Characterization of the 1918 'Spanish' Influenza Virus."

86 *Johan Hultin came to the United States:* Many of the details are from Kolata, *Flu*, chapter 4, and David Brown, "Resurrecting 1918 Flu Virus Took Many Turns," *Washington Post*, October 10, 2005.

86 *"try to find a victim of the 1918 flu pandemic":* From an interview with Hultin on Nova: "The 1918 Flu Pandemic That Infected 500 Million People."

86 *Otto Geist:* Kolata, *Flu*, 95–98.

87 *he was given permission to proceed:* Ibid., 108.

88 *they found three other bodies:* Hultin, in the Nova documentary.

88 *Hultin's expedition remained forgotten:* Kolata notes that Hultin never published the results of his efforts (*Flu*, 115). There was a brief newspaper report of the expedition in the *Washington Post*: N. S. Haseltine, "Scientists Seek 1918 Flu Virus," September 2, 1951.

88 *"I'm afraid the warranty will run out":* Elizabeth Fernandez, "The Virus Detective / Dr. John Hultin Has Found Evidence of the 1918 Flu Epidemic That Had Eluded Experts for Decades," SFGate. Accessed May 2, 2018. https://www.sfgate.com/magazine/article/The-Virus-detective-Dr-John-Hultin-has-found-2872017.php.

90 *paper that detailed the findings of Private Roscoe's flu virus:* Taubenberger et al., "Initial Genetic Characterization of the 1918 'Spanish' Influenza Virus."

90 *Taubenberger offered to analyze:* According to Kolata, Taubenberger resigned from the Duncan team after it was alleged that it was charging for media interviews. See also Kirsty Duncan, *Hunting the 1918 Flu: One Scientist's Search for a Killer Virus* (Toronto: University of Toronto Press, 2003), 65–85.

90 *buoyed by a federal grant for $150,000:* Kolata, *Flu*, 272.

91 *the lab confirmed the presence of 1918 flu particles:* Ibid., 262–65.

91 *now rebuild the entire genetic code of the 1918 virus:* Taubenberger, Reid, and Fanning, "Capturing a Killer Flu Virus."

91 *the* New York Times *reported:* John Noble, "Quest for Frozen Pandemic Virus Yields Mixed Results," *New York Times*, September 7, 1998, F3.

91 *she found renown later:* Duncan went on to write a book about her experiences and on becoming Canada's minister of science. She has been referred

to as a "badass." See Maryn McKenna, "Canada's First (and Female) Science Minister Is a Badass," *National Geographic*. Accessed April 26, 2018. http://phenomena.nationalgeographic.com/2015/11/05/canadas-first-and-female-science-minister-is-a-badass/. Her expedition is discussed in detail in Kolata, *Flu*, chapter 9, and in even greater depth in Duncan, *Hunting the 1918 Flu*, from where I have drawn the story.

92 *The search widened:* There were other attempts to exhume the virus. In September 2008, the British virologist John Oxford led a team that dug up the body of Sir Tatton Benvenuto Mark Sykes. Sykes had risen through the ranks of the British Army to become a colonel, and during World War I, in anticipation of the collapse of the Ottoman Empire, he helped to carve up the Middle East for European powers. In February 1919 Sykes died at the age of thirty-nine, a victim of the tail end of the flu pandemic. He was laid to rest in a lead coffin in St. Mary's Church in Yorkshire, next to Sledmere House, the ancestral home of the Sykes family. (Sledmere, like so many of Britain's aristocratic homes, is now making ends meet by providing tours; it is also available for weddings.) Lead coffins seal off the body from the surrounding earth. This slows the natural decay of the body, and so Sykes was a prime candidate for exhumation. John Oxford was hopeful, and became the subject of a documentary about the project. "We're on the verge of the first influenza pandemic of the twenty-first century," he told the BBC, "and we think Sir Mark can help us." Sir Mark's descendants agreed. "It is rather fascinating that maybe even in his state as a corpse, he might be helping the world in some way," his grandson said.

Digging up the remains of British aristocracy required some paperwork, which took about two years to complete. Oxford had to obtain permission from the church court covering the Diocese of York, the Department for Constitutional Affairs, and the Health and Safety Executive. A short prayer was offered at the graveside, and then scientists wearing hazmat suits and oxygen masks went to work. They were disappointed and frustrated when, after reaching the coffin, they discovered that its cover was not intact. The body had decayed. After testing seventeen samples of the remains of Sir Mark, Oxford and his team failed to find the virus. The legacy of Sir Mark Sykes would remain tied entirely to his 1916 plan for the Middle East, and not to the 1918 virus that killed him. See James Barr, *A Line in the Sand: France and the Struggle for the Mastery of the Middle East* (New York: Simon & Schuster, 2011), and "Vital Flu Clue," BBC. Accessed April 26, 2018. http://www.bbc.co.uk/insideout/yorkslincs/series11/week8_flu.shtml.

92 *died due to a secondary bacterial infection:* A. H. Reid et al., "1918 Influenza

Pandemic Caused by Highly Conserved Viruses with Two Receptor-Binding Variants," *Emerging Infectious Diseases* 9, no. 10 (2003): 1249–53.

92   *the genetic fingerprints across all the samples:* Ibid.

92   *a single strain of influenza was in circulation in the early stages:* Ibid.

93   *In 2005 the team announced:* T. M. Tumpey et al., "Characterization of the Reconstructed 1918 Spanish Influenza Pandemic Virus," *Science* 310, no. 5745 (2005): 77–80.

94   *the 1918 virus appeared to be a bird virus:* J. K. Taubenberger et al., "Characterization of the 1918 Influenza Virus Polymerase Genes," *Nature* 437, no. 7060 (2005): 889–93.

95   *October 2005 issue of* Science: Tumpey et al., "Characterization of the Reconstructed 1918 Spanish Influenza Pandemic Virus," *Science* 310, no. 5745 (2005): 77–80.

95   *Scientists from Stony Brook University:* J. Cello, A. V. Paul, and E. Wimmer, "Chemical Synthesis of Poliovirus cDNA: Generation of Infectious Virus in the Absence of Natural Template," *Science* 297, no. 5583 (2002): 1016–18.

95   Biotechnology Research in an Age of Terrorism: National Research Council, *Biotechnology Research in an Age of Terrorism* (Washington, D.C.: The National Academies, 2004).

96   *Kennedy stuck by his decision:* D. Kennedy, "Better Never Than Late," *Science* 310, no. 5746 (2005): 195. Some scientists were not convinced by reassurances that it was safe to work with the active 1918 virus, and felt that the benefits of what we could learn were outweighed by the risks of bringing the virus back from the dead. Richard Ebright, a microbiologist at Rutgers University, believed that the research was too dangerous to have been performed. "If the virus was to be accidentally or intentionally released," he said, "it is virtually certain that there would be greater lethality than from seasonal influenza, and quite possible that the threat of pandemic that is in the news daily would become a reality." But nobody had asked Professor Ebright, who made his comments three months after the announcement that the 1918 virus had been rebuilt. See Jamie Shreeve, "Why Revive a Deadly Virus?," *New York Times Magazine*, January 29, 2006, 48.

96   *In 2012 an international group genetically modified:* S. Herfst et al., "Airborne Transmission of Influenza A/H5NI Virus between Ferrets," *Science* 336, no. 6088 (2012): 1534–41.

97   *more lethal than the original avian virus:* T. Watanabe et al., "Circulating Avian Influenza Viruses Closely Related to the 1918 Virus Have Pandemic Potential," *Cell Host & Microbe* 15, no. 6 (2014): 692–705.

97   *the White House paused the federal funding:* "Doing Diligence to Assess the

Risks and Benefits of Life Sciences Gain-of-Function Research," The White House. Accessed April 26, 2018. https://obamawhitehouse.archives.gov/blog/2014/10/17/doing-diligence-assess-risks-and-benefits-life-sciences-gain-function-research.

97   *to assess the risks and benefits:* "U.S. Government Gain-of-Function Deliberative Process and Research Funding Pause on Selected Gain-of-Function Research Involving Influenza, MERS, and SARS Viruses," n.p. n.p October 17, 2014. Available at https://www.phe.gov/s3/dualuse/Documents/gain-of-function.pdf.

97   *"I shall sleep better tonight":* Donald McNeil, "White House to Cut Funding for Risky Biological Study," *New York Times,* October 17, 2014.

97   *the White House released new research guidelines:* "Recommended Policy Guidance for Departmental Development of Review Mechanisms for Potential Pandemic Pathogen Care and Oversight (P3CO)," n.p, n.p. Available at https://www.phe.gov/s3/dualuse/Documents/P3CO-FinalGuidanceStatement.pdf.

98   *the NIH promptly removed its ban:* "NIH Lifts Funding Pause on Gain-of-Function Research," The NIH Director. Accessed May 2, 2018. https://www.nih.gov/about-nih/who-we-are/nih-director/statements/nih-lifts-funding-pause-gain-function-research.

98   *"I know nothing":* Interview with Jeffrey Taubenberger, June 14, 2016.

6. DATA, INTUITION, AND OTHER WEAPONS OF WAR

101   *in sunny Florida only 43 providers take part:* This number appears in the *Florida Flu Review,* 2015–2016 season, Week 14: April 3–9, 2016.

103   *flu queries spiked on Google Flu Trends:* Google.org, "Google Flu Trends Overview," YouTube video. Accessed November 6, 2017. https://www.youtube.com/watch?v=6111nS66Dpk.

104   *Canada:* M. T. Malik et al., " 'Google Flu Trends' and Emergency Department Triage Data Predicted the 2009 Pandemic H1N1 Waves in Manitoba," *Canadian Journal of Public Health* 102, no. 4 (2011): 294–97.

104   *Australia:* H. Kelly and K. Grant, "Interim Analysis of Pandemic Influenza (H1N1) 2009 in Australia: Surveillance Trends, Age of Infection and Effectiveness of Seasonal Vaccination," *Eurosurveillance* 14, no. 31 (2009).

104   *several European countries:* A. Valdivia et al., "Monitoring Influenza Activity in Europe with Google Flu Trends: Comparison with the Findings of Sentinel Physician Networks—Results for 2009–10," *Eurosurveillance* 15, no. 29 (2010).

104 *the sale of antiviral medications:* A. Patwardhan and R. Bilkovski, "Comparison: Flu Prescription Sales Data from a Retail Pharmacy in the U.S. with Google Flu Trends and U.S. ILINet (CDC) Data as Flu Activity Indicator," *PLoS One* 7, no. 8 (2012): e43611.

104 *The algorithm was updated:* S. Cook et al., "Assessing Google Flu Trends Performance in the United States during the 2009 Influenza Virus A (H1N1) Pandemic," *PLoS One* 6, no. 8 (2011): e23610.

104 *it turned out that Google had overestimated:* D. Lazer et al., "The Parable of Google Flu: Traps in Big Data Analysis," *Science* 343, no. 6176 (2014): 1203–5.

105 *In the influential journal* Science: Perhaps, as one group of researchers put it, "the algorithm producing the data (and thus user utilization) has been modified by the service provider in accordance with their business model." See ibid., 1204.

105 *New York declared a public health emergency:* "Governor Cuomo Declares State Public Health Emergency in Response to Severe Flu Season," New York State. Accessed April 26, 2018. https://www.governor.ny.gov/news/governor -cuomo-declares-state-public-health-emergency-response-severe-flu-season. See also Tim Hartford, "Big Data: Are We Making a Big Mistake?," *Significance*, December 2014.

105 *Boston's Street Bump app:* Ibid.

106 *said Alain-Jacques Valleron:* Cited in D. Butler, "When Google Got Flu Wrong," *Nature* 494, no. 7436 (2013): 155–56.

106 *a goodbye note of sorts:* The Flu Trends Team, "The Next Chapter for Flu Trends," Google Research Blog. Accessed May 2, 2018. https://research .googleblog.com/2015/08/the-next-chapter-for-flu-trends.html.

107 *Nelson started modestly:* Telephone interview with Forrest Nelson, November 16, 2017.

107 *It correctly forecast:* P. M. Polgreen, F. D. Nelson, and G. R. Neumann, "Use of Prediction Markets to Forecast Infectious Disease Activity," *Clinical Infectious Diseases* 44, no. 2 (2007): 272–79.

108 *I spoke with Nelson:* Telephone interview with Forrest Nelson, November 16, 2017.

111 *The program started by receiving weekly reports:* D. Das et al., "Monitoring Over-the-Counter Medication Sales for Early Detection of Disease Outbreaks—New York City," *MMWR Supplements* 54 (2005): 4–46.

112 *Maryland Resident Influenza Tracking Survey:* The number of participants is taken from email correspondence with Stephen Stanley of the Maryland Department of Health, April 27, 2018. See also "Maryland Influenza Sur-

veillance Report: 2008–09 Influenza Season Summary," Division of Communicable Disease Surveillance, Office of Epidemiology and Disease Control Programs, Maryland Department of Health and Mental Hygiene, 2009.

112 *2014–2015 Maryland influenza season:* "Influenza in Maryland. 2014–2015 Season Report," Division of Communicable Disease Surveillance, Office of Epidemiology and Disease Control Programs, Maryland Department of Health and Mental Hygiene, 2014.

113 *Sharon Sanders is the editor in chief:* The quotes and information about the history and formation of FluTrackers comes from a phone interview with Sanders, April 24, 2017, and from follow-up email correspondence.

113 *President George W. Bush was reading a history of the flu:* Warren Vieth, "Bush Salts His Summer with Eclectic Reading List," *Los Angeles Times,* August 16, 2005. Available at http://articles.latimes.com/2005/aug/16/nation/na-bushread16. Accessed May 2, 2018.

115 *In an email:* Email dated November 16, 2017.

7. YOUR EVENING FLU FORECAST

119 *the correlation between the phases of the moon and visits to the emergency room:* See, for example, S. Kamat et al., "Pediatric Psychiatric Emergency Department Visits during a Full Moon," *Pediatric Emergency Care* 30, no. 12 (2014): 875–78.

120 *Other infectious diseases have a seasonality:* M. Oshinsky, *Polio: An American Story* (Oxford: Oxford University Press, 2005), 9–10; N. B. Mantilla-Beniers et al., "Decreasing Stochasticity Through Enhanced Seasonality in Measles Epidemics," *Journal of the Royal Society Interface* 7, no. 46 (2009): 727–39.

120 *"It should not cause any greater importance":* "The Epidemic of Influenza," *JAMA* 71, no. 13 (1918): 1063–64. While it is true that there were various waves of flu during the pandemic, the overall trends of the disease generally followed the usual pattern: a rise in cases over the fall and winter.

120 *that the way in which air flows:* G. W. Hammond, R. L. Raddatz, and D. E. Gelskey, "Impact of Atmospheric Dispersion and Transport of Viral Aerosols on the Epidemiology of Influenza," *Reviews of Infectious Diseases* 11, no. 3 (1989): 494–97. The authors hypothesized that "the long-range atmospheric transport of aerosolized influenza . . . and the seasonal changes in atmospheric circulation patterns may lead to the regular annual cycles of influenza activity." This hypothesis was offered as one model "intended to contribute ideas for discussion." It's almost as if the authors didn't quite manage to convince themselves.

NOTES

120 *In the tropics:* S. Hirve et al., "Influenza Seasonality in the Tropics and Subtropics—When to Vaccinate?," *PLoS One* 11, no. 4 (2016): e0153003.

121 *when you dig deeper you find many problems:* J. J. Cannell et al., "Epidemic Influenza and Vitamin D," *Epidemiology & Infection* 134, no. 6 (2006): 1129–40.

121 *we tend to use public transportation more:* Y. Yang, A. V. Diez Roux, and C. R. Bingham, "Variability and Seasonality of Active Transportation in USA: Evidence from the 2001 NHTS," *International Journal of Behavioral Nutrition and Physical Activity* 8 (2011): 96. The reasons are not clear, but may have something to do with our use of public transport for vacations.

121 *"London Transport would ensure an all-the-year epidemic":* C. Andrews, *The Common Cold* (New York: W.W. Norton, 1965), 137.

122 *arrived by comet, as microscopic alien hitchhikers:* Fred Hoyle and Chandra Wickramasinghe, *Evolution from Space: A Theory of Cosmic Creationism* (New York: Simon & Schuster, 1982).

122 *believed instead in a steady-state universe:* Simon Mitton, *Fred Hoyle: A Life in Science*, pbk. ed. (Cambridge: Cambridge University Press, 2011).

122 *letter in the prestigious scientific journal* Nature: Fred Hoyle and N. C. Wickramasinghe, "Sunspots and Influenza," *Nature* 343, no. 6256 (1990): 304. Hoyle admitted that he was not the first to notice a relationship between solar activity and flu outbreaks. That accolade went to another Brit, by the name of Robert Hope-Simpson. See R. E. Hope-Simpson, "Sunspots and flu: a correlation," *Nature* 265, no. 5676 (1978): 86.

122 *in the words of NASA:* "Impacts of Strong Solar Flares," NASA. Accessed April 29, 2018. https://www.nasa.gov/mission_pages/sunearth/news/flare-impacts.html.

122 *it can be manipulated to fit any model:* A. von Alvensleben, "Influenza according to Hoyle," *Nature* 344 (1990): 374.

123 *rays transform it into vitamin D:* For this discovery, the German chemist Adolf Windaus was awarded a Nobel Prize in 1928.

123 *twice as many people die over the dark winter:* Cannell et al., "Epidemic Influenza and Vitamin D."

124 *Britain's senior citizens:* V. Hirani and P. Primatesta, "Vitamin D Concentrations among People Aged 65 Years and Over Living in Private Households and Institutions in England: Population Survey," *Age and Ageing* 34 (2006): 485–91.

124 *also much more common in African Americans:* A. Zadshir et al., "The Prevalence of Hypovitaminosis D among U.S. Adults: Data from the NHANES III," *Ethnicity & Disease* 15, no. 4, suppl. 5 (2005): s5–97–101.

124 *their mortality from pneumonia and influenza:* Cannell et al., "Epidemic Influenza and Vitamin D."

124 *the winter group was eight times more likely to develop a fever:* A. S. Shadrin, I. G. Marinich, and L. Y. Taros, "Experimental and Epidemiological Estimation of Seasonal and Climato-geographical Features of Non-specific Resistance of the Organism to Influenza," *Journal of Hygiene, Epidemiology, Microbiology, and Immunology* 21, no. 2 (1977): 155–61.

124 *a group of schoolchildren in Japan:* M. Urashima et al., "Randomized Trial of Vitamin D Supplementation to Prevent Seasonal Influenza A in Schoolchildren," *American Journal of Clinical Nutrition* 91, no. 5 (2010): 1255–60.

124 *healthy adults in New Zealand:* D. R. Murdoch et al., "Effect of Vitamin D_3 Supplementation on Upper Respiratory Tract Infections in Healthy Adults: The Vidaris Randomized Controlled Trial," *JAMA* 308, no. 13 (2012): 1333–39. An accompanying editorial called for vitamin D to join the list of treatments that were "ineffective for preventing or treating upper respiratory tract infections in healthy adults." That list includes "echinacea, zinc, steam inhalation, vitamin C, garlic, antihistamines, Chinese medicinal herbs, intranasal corticosteroids, intranasal ipratroprium, Pelargonium sidoides herbal extract, [and] saline nasal irrigation . . ." See J. A. Linder, "Vitamin D and the Cure for the Common Cold," *JAMA* 308, no. 13 (2012): 1375–76.

124 *when older adults took extra doses of vitamin D:* Abhimanyu and A. K. Coussens, "The Role of UV Radiation and Vitamin D in the Seasonality and Outcomes of Infectious Disease," *Photochemical and Photobiological Sciences* 16, no. 3 (2017): 314–38. It is not just Britain's elderly. A 2010 study from the Institute of Medicine noted that "sun exposure currently contributes meaningful amounts of vitamin D to North Americans and [the data] indicates that a majority of the population is meeting its needs for vitamin D. Nonetheless, some subgroups—particularly those who are older and living in institutions or who have dark skin pigmentation—may be at increased risk for getting too little vitamin D." See their report brief, "Dietary Reference Intakes for Calcium and Vitamin D," Institute of Medicine, November 2010. http://www.nationalacademies.org/hmd/~/media /Files/Report%20Files/2010/Dietary-Reference-Intakes-for-Calcium -and-Vitamin-D/Vitamin%20D%20and%20Calcium%202010%20Report %20Brief.pdf.

125 *One such analysis pooled the results:* P. Bergman et al., "Vitamin D and Respiratory Tract Infections: A Systematic Review and Meta-analysis of Randomized Controlled Trials," *PLoS One* 8, no. 6 (2013): e65835.

125 *a group from New York's Mount Sinai School of Medicine:* A. C. Lowen et al., "Influenza Virus Transmission Is Dependent on Relative Humidity and Temperature," *PLoS Pathogens* 3, no. 10 (2007): 1470–76.

125 *varying both the temperature and the humidity:* There are two measures of humidity: relative and absolute. Relative humidity is what you hear about on the weather forecast. It measures how much water is in the atmosphere compared to the amount needed to saturate the particular environment that's being measured. When the weather forecasters tell you about the outside humidity, this is what they mean. If the amount of water in the atmosphere stays the same, the humidity goes down as the temperature goes up. In winter, the air is colder, so it can hold less water. As a result, the relative humidity is higher in the winter. Shaman and his group at Columbia focused on absolute humidity, which is a measure of the amount of water in a particular atmosphere, period, without regard to another variable. By focusing on the absolute humidity, researchers could eliminate temperature as a variable. It also better reflects the way in which indoor and outdoor humidity are related. Inside our homes, absolute humidity correlates very well with absolute humidity outside, whereas inside and outside relative humidity are poorly correlated. See J. L. Nguyen, J. Schwartz, and D. W. Dockery, "The Relationship between Indoor and Outdoor Temperature, Apparent Temperature, Relative Humidity, and Absolute Humidity," *Indoor Air* 24, no. 1 (2014): 103–12. A note of caution, though: the sample size of this study was only sixteen homes.

126 *The uninfected animals remained happy and healthy:* Guinea pigs were also victims of the 1918 influenza epidemic. In September 1918 at Camp Cody, a military base in New Mexico, there was an outbreak of influenza that lasted three months. By the time it was over, the base hospital had admitted over 3,000 patients, more than a quarter of all the troops stationed there. Five nurses and almost 250 soldiers had died. In a medical report about the outbreak, the authors described the temperature, blood counts, and urine analysis of the infected soldiers, and then they took an unexpected detour. They noted how, shortly after the 1918 epidemic reached the camp, the guinea pigs in their lab began to die. At first, the physicians thought that they had died from food poisoning, but at postmortem they found "the unmistakable signs of pneumonia." The animals suffered in the same way that their military cocampers had. "During this time," they wrote, "the sick animal trembling from chills, with hair ruffled, sat huddled up in a corner of the pen, moving about only to eat. This it would do until shortly before death. The respirations were rapid and wheezing; the characteristic shrill whistle became scarcely audible. The animal was apparently in a stupor which gradually deepened until death supervened. . . . Just

before death the animal would fall on one side, rise a time or two, then make a few feeble efforts to do so again. Within fifteen or thirty minutes it would die." See Frederick Lamb and Edward Brannin, "The Epidemic Respiratory Infection at Camp Cody N.M.," *JAMA* 72, no. 15 (1919): 1056–62.

126 *they created a simulated coughing machine:* J. D. Noti et al., "High Humidity Leads to Loss of Infectious Influenza Virus from Simulated Coughs," *PLoS One* 8, no. 2 (2013): e57485.

127 *"infectious disease dynamics are nonlinear":* J. Shaman and A. Karspeck, "Forecasting Seasonal Outbreaks of Influenza," *Proceedings of the National Academies of Science* 109, no. 50 (2012): 20425–30.

127 *ensemble forecasting:* Ibid.

128 *their weather model was about 75 percent accurate:* J. Shaman et al., "Real-Time Influenza Forecasts during the 2012–2013 Season," *Nature Communications* 4 (2013): 2837.

128 *Predict the Influenza Season Challenge:* "CDC Announces Winner of the 'Predict the Influenza Season Challenge,'" Centers for Disease Control and Prevention. Accessed April 20, 2017. https://www.cdc.gov/flu/news/predict-flu-challenge-winner.htm.

128 *Almost half of the existing ventilators:* This data comes from X. Zhang, M. I. Meltzer, and P. M. Wortley, "FluSurge—a Tool to Estimate Demand for Hospital Services During the Next Pandemic Influenza," *Medical Decision Making* 26, no. 6 (2006): 617–23. It assumes a pandemic lasting eight weeks that attacks 25 percent of the population. The numbers are even worse if the pandemic lasts three months and infects a third of the population. In that scenario, 85 percent of the ventilators would be needed for the sickest of the flu patients. The authors were from the CDC in Atlanta, which is probably why they modeled their numbers on that city. The tool they created to produce these estimates is called FluSurge and is openly available. But be warned: the numbers are scary.

129 *hospitals are short of nurses:* There are many academic reports about a nursing shortage and they are not consistent. The Bureau of Labor Statistics noted a recent increase in the number of nurses entering the market, which has led to competition for jobs. But only in some places. "Registered Nurses," United States Department of Labor. Accessed April 29 2017. https://www.bls.gov/ooh/healthcare/registered-nurses.htm#tab-6.

129 *the rate of vaccination in the United States:* Tammy A. Santibanez et al., "Flu Vaccination Coverage, United States, 2015–16 Influenza Season," Centers for Disease Control and Prevention. Accessed April 29, 2018. https://www.cdc.gov/flu/fluvaxview/coverage-1516estimates.htm.

129 *Communities exposed to warnings:* Kristin Dow and Susan Cutter, "Crying Wolf: Repeat Responses to Hurricane Evacuation Orders," *Coastal Management* 26, no. 4 (1998): 237–52. This paper discussed the reaction to gubernatorial warnings, which, like those from public health offices, are governmental.

8. THE FAULT IN OUR STOCKPILES: TAMIFLU AND THE CURE THAT WASN'T THERE

131 *The bunkers contain our Strategic National Stockpile:* I asked to visit one of these bunkers, but was turned down, apparently as a result of "a change in policy." I blame Nell Greenfieldboyce, a reporter for National Public Radio, who was the first—and apparently last—reporter to visit the stockpile. "Since I had to sign a confidentiality agreement," she said, "I can't describe the outside. But the inside is huge." See "Inside a Secret Government Warehouse Prepped for Health Catastrophes," *National Public Radio*, June 27, 2016. http://www.npr.org/sections/health-shots/2016/06/27/483069862/inside-a-secret-government-warehouse-prepped-for-health-catastrophes.

132 *"A cross between Amazon and a local pharmacy"*: Interview with Mark Burchess, November 28, 2016.

134 *120 million doses of the H1N1 vaccine were shipped:* "2009 H1N1 Flu," Centers for Disease Control and Prevention. Accessed April 29, 2018. https://www.cdc.gov/h1n1flu/vaccination/vaccinesupply.htm.

134 *baby flu particles could not spread the infection:* A. Moscona, "Neuraminidase Inhibitors for Influenza," *New England Journal of Medicine* 353, no. 13 (2005): 1363–73.

135 *a group of researchers from Scotland:* The authors were a pessimistic bunch, and were less than enthusiastic about their discovery. While noting that it was *possible* that it could lead to an antiviral drug, they noted the drugs appeared to be "too labile biologically to give any anti-influenza effect in an intact animal." See J. D. Edmond et al., "The Inhibition of Neuraminidase and Antiviral Action," *British Journal of Pharmacology and Chemotherapy* 27, no. 2 (1966): 415–26.

135 *scientists began to test them:* Walter Sneader, *Drug Discovery: A History* (Hoboken, NJ: Wiley & Sons, 2005), 264–65; F. G. Hayden et al., "Safety and Efficacy of the Neuraminidase Inhibitor GG167 in Experimental Human Influenza," *JAMA* 275, no. 4 (1996): 295–99.

135 *a TV ad for Tamiflu:* "Tamiflu TV Commercial, 'Small House,'" iSpot.tv. Accessed April 29, 2018. https://www.ispot.tv/ad/77Nb/tamiflu-small-house.

135 *targeted mothers whose children had the flu:* "Tamiflu TV Commercial, 'Kids,'" iSpot.tv. Accessed April 29, 2018. https://www.ispot.tv/ad/AIqV/ tamiflu-kids. This ad aired almost three thousand times.

136 *Two years after the outbreak the WHO published a report:* World Health Organization Department of Communicable Disease Surveillance and Response, "Influenza Pandemic Plan. The Role of WHO and Guidelines for National and Regional Planning" (Geneva, 1999). The quotation is on p. 54.

136 *funded by at least seven pharmaceutical companies:* This list appears on the ESWI website: "Resources," European Scientific Working Group on Influenza. Accessed April 29, 2018. http://eswi.org/home/about-eswi/resources/.

136 *These manufacturers joined together:* Ibid. According to the *British Medical Journal*, the ESWI is a group "funded entirely by Roche and other influenza manufacturers." (See D. Cohen and P. Carter, "WHO and The Pandemic Flu 'Conspiracies,'" *British Medical Journal* 340 (2010): c2912.) This may have been true in years past, but the ESWI currently lists public funding as well as "unrestricted grants by vaccine and antivirals companies."

136 *the 1997 avian influenza outbreak:* The WHO report mentioned that two other antiviral drugs (amantadine and its derivative rimantadine) had "been shown to be clinically effective in preventing illness" and that they could "reduce the severity and duration of illness when taken early after onset." These two drugs were not neuraminidase inhibitors, but seemed to work against the flu virus. Since then the clever flu virus has become resistant to them to such a degree that they are no longer given for the treatment of influenza.

137 *In 1999 the Cochrane Collaboration:* T. Jefferson et al., "Neuraminidase Inhibitors for Preventing and Treating Influenza in Healthy Adults," *Cochrane Database of Systematic Reviews* 2 (1999).

137 *Cochrane reviewers pledge to issue reports:* Cochrane, "About us." Accessed April 29, 2018. http://www.cochrane.org/about-us.

137 *a warning letter about its advertising campaign:* Department of Health and Human Services, FDA letter MACMIS ID#8675, April 14, 2000. https://www .fda.gov/downloads/Drugs/GuidanceComplianceRegulatoryInformation /EnforcementActivitiesbyFDA/WarningLettersandNoticeofViolationLetters toPharmaceuticalCompanies/UCM166329.pdf.

137 *His speech there:* "President Outlines Pandemic Influenza Preparations and Response," The White House. Accessed April 29, 2018. https://georgewbush-whitehouse.archives.gov/news/releases/2005/11/20051101-1.html.

138 *an effort to detect flu outbreaks earlier:* To do this, there would be an inter-

national effort to be led by the recently formed International Partnership on Avian and Pandemic Influenza. Within the U.S., the president launched the National Biosurveillance Initiative, which somehow "would provide continual situational awareness." Ibid.

138 *the president requested $7.1 billion:* T. Salaam-Blyther, "U.S. and International Responses to the Global Spread of Avian Flu: Issues for Congress," in *Congressional Research Service Report for Congress* (Washington, D.C.: Congressional Research Service, 2006), 7.

138 *five times as much Tamiflu was being prescribed:* J. R. Ortiz et al., "Oseltamivir Prescribing in Pharmacy-Benefits Database, United States, 2004–2005," *Emerging Infectious Diseases* 14, no. 8 (2008): 1280–83.

138 *healthy people were buying up Tamiflu:* A. S. Brett and A. Zuger, "The Run on Tamiflu—Should Physicians Prescribe on Demand?," *New England Journal of Medicine* 353, no. 25 (2005): 2636–37.

139 *threats of a shortage:* D. Spurgeon, "Roche Canada Stops Distributing Oseltamivir," *British Medical Journal* 331, no. 7524 (2005): 1041.

139 *"we will be unable to prevent it reaching the UK":* "Britain Reveals Flu Pandemic Plan," *BBC News*, March 1, 2005. Accessed April 29, 2018. http://news .bbc.co.uk/2/hi/health/4305813.stm. The BBC also reported on the government's decision to stockpile Tamiflu.

139 *only 2.3 million doses were on hand:* U.S. Department of Health and Human Services, "HHS Pandemic Influenza Plan" (2005), F-39.

139 *another analysis of antiflu medications:* T. Jefferson et al., "Antivirals for Influenza in Healthy Adults: Systematic Review," *Lancet* 367, no. 9507 (2006): 303–13.

140 *The World Health Organization declared a pandemic:* Margaret Chan, "World Now at the Start of 2009 Influenza Pandemic," World Health Organization. Accessed April 29, 2018. http://www.who.int/mediacentre/news/statements /2009/h1n1_pandemic_phase6_20090611/en/. This episode is discussed in greater detail in chapter 4.

140 *not as virulent as experts feared:* Still, the 2009 pandemic might have caused additional deaths in the U.S. when compared to the seasonal influenza. See Shrestha et al., "Estimating the Burden of 2009 Pandemic Influenza A."

140 *Ropes & Gray:* Liz Kowalczyk, "Firms' Deals for Flu Drug Draw Fire," *Boston Globe*, October 30, 2009.

140 *the Boston Globe admonished Ropes & Gray:* "Swine Flu: Firms Shouldn't Hoard Drugs," *Boston Globe*, November 3, 2009.

140 *The CDC issued a curt statement:* Kowalczyk, "Firms' Deals for Flu Drug Draw Fire."

141 *Dr. Karen Victor:* Ibid.

141 *the flu virus was 100 percent resistant to Tamiflu:* Centers for Disease Control and Prevention, "Update: Drug Susceptibility of Swine-Origin Influenza A (H1N1) Viruses, April 2009," *Morbidity and Mortality Weekly Report* 58, no. 16 (2009): 435–55.

141 *the U.S., Britain, and at least ninety-four other countries:* A. Jack, "Flu's Unexpected Bonus," *British Medical Journal* 339 (2009): b3811.

141 *The review by Dr. Kaiser and her colleagues:* L. Kaiser et al., "Impact of Oseltamivir Treatment on Influenza-Related Lower Respiratory Tract Complications and Hospitalizations," *Archives of Internal Medicine* 163, no. 14 (2003): 1667–72.

143 *In December 2009 Jefferson released an updated review:* T. Jefferson, M. Jones, P. Doshi, and C. Del Mar, "Neuraminidase inhibitors for preventing and treating influenza in healthy adults: systematic review and meta-analysis," *British Medical Journal* 339 (2009): b5106.

143 *Paul Flynn sponsored a motion:* "Early day motion 669," www.parliament.uk. Accessed April 29, 2018. http://www.parliament.uk/edm/print/2009-10/669. For some odd reason, the motion described the distribution of Tamiflu as a *vaccination* program. It wasn't.

143 *salting Britain's snowy roads:* Z. Kmietowicz, "Use Leftover Tamiflu to Grit Icy Roads, MP Suggests," *British Medical Journal* 340 (2010): c501.

143 *the way the swine flu pandemic had been handled:* "The Handling of the H1N1 Pandemic: More Transparency Needed," Social, Health and Family Affairs Committee, Council of Europe. Accessed April 29, 2018. http://assembly .coe.int/CommitteeDocs/2010/20100329_MemorandumPandemie_E.pdf.

144 *Fiona Godlee, the editor in chief:* F. Godlee, "Conflicts of Interest and Pandemic Flu," *British Medical Journal* 340 (2010): c2947. Godlee cited an in-depth review of WHO influenza experts and their ties to industry. See D. Cohen, "WHO and the Pandemic Flu 'Conspiracies,'" *British Medical Journal* 340 (2010): c2912.

144 *the review showed that Tamiflu reduced some of the complications:* M. A. Hernan and M. Lipsitch, "Oseltamivir and Risk of Lower Respiratory Tract Complications in Patients with Flu Symptoms: A Meta-analysis of Eleven Randomized Clinical Trials," *Clinical Infectious Diseases* 53, no. 3 (2011): 277–79.

144 *Roche released all the trials requested by the Cochrane group:* T. Jefferson and P. Doshi, "Multisystem Failure: The Story of Anti-Influenza Drugs," *British Medical Journal* 348 (2014): g2263.

144 *which it released in April 2014:* T. Jefferson et al., "Neuraminidase Inhibi-

tors for Preventing and Treating Influenza in Healthy Adults and Children," *Cochrane Database of Systematic Reviews* 4 (2014): CD008965. In addition, the group reviewed more than 160,000 pages of regulatory documents.

145 *Tamiflu did not reduce the risk of hospitalization:* Footnote for the health policy wonks: the Cochrane review wrote on page 2 that "Oseltamivir significantly reduced self reported, investigator-mediated, unverified pneumonia. . . . The effect was not significant in the five trials that used a more detailed diagnostic form for pneumonia. There were no definitions of pneumonia (or other complications) in any trial. No oseltamivir treatment studies reported effects on radiologically confirmed pneumonia. There was no significant effect on unverified pneumonia in children."

145 *the* Lancet *released an analysis:* J. Dobson et al., "Oseltamivir Treatment for Influenza in Adults: A Meta-analysis of Randomised Controlled Trials," *Lancet* 385, no. 9979 (2015): 1729–37.

145 *recommended antiviral drugs only for those high-risk patients:* See, for example, "What You Should Know About Influenza (Flu) Antiviral Drugs," Centers for Disease Control and Prevention. Accessed April 29, 2018. https://www .cdc.gov/flu/pdf/freeresources/updated/antiviral-factsheet-updated.pdf. See also "Influenza Antiviral Medications: Summary for Clinicians," Centers for Disease Control and Prevention. Accessed April 29, 2018. https://www.cdc .gov/flu/professionals/antivirals/summary-clinicians.htm.

145 *At a telephone news briefing:* "Transcript for CDC Telebriefing: Update on Flu Season 2014–15," Centers for Disease Control and Prevention. Accessed April 29, 2018. https://www.cdc.gov/media/releases/2014/t1204-flu-season.html.

146 *earned his doctorate from MIT:* P. Doshi, "Influenza: A Study of Contemporary Medical Politics" (Massachusetts Institute of Technology, 2011).

146 *We met there:* Interview on June 5, 2017.

147 *the Department of Health and Human Services believed:* U.S. Department of Health and Human Services, "HHS Pandemic Influenza Plan," n.d., Appendix, D20. Much of the information in the following paragraphs is based on a telephone interview with Peter Palese, October 20, 2017.

149 *On the front page was an article:* D. Runde, "Still Prescribing Oseltamivir?," *Emergency Medicine News* 39, no. 4. (2017): 1, 41.

9. THE HUNT FOR A FLU VACCINE

151 *Vaccination:* The terms "vaccination" and "inoculation" are used synonymously.

151 *inoculation was used by the Chinese:* C. P. Gross and K. A. Sepkowitz, "The

Myth of the Medical Breakthrough: Smallpox, Vaccination, and Jenner Reconsidered," *International Journal of Infectious Disease* 3, no. 1 (1998): 54–50.

151 *priests traveled the Indian countryside:* Neils Brimnes, "Variolation, Vaccination and Popular Resistance in Early Colonial South India," *Medical History* 48, no. 2 (2004): 199–228.

151 *Edward Jenner, a British physician:* Biographic information is from S. Riedel, "Edward Jenner and the History of Smallpox and Vaccination," *Proceedings (Baylor University Medical Center)* 18, no. 1 (2005): 21–25.

153 *pleaded for his colleagues to stop bickering:* E. Rosenow, "Prophylactic Inoculation against Respiratory Infections," *JAMA* 72, no. 1 (1919): 31–34.

153 *Leary mixed these samples together:* Timothy Leary, "The Use of the Influenza Vaccine in the Present Epidemic," *American Journal of Public Health* 8, no. 10 (1918): 754–55.

153 *where at least 18,000 people were inoculated:* Crosby, *America's Forgotten Pandemic*, 100.

153 *"fluphobia":* Price, "Influenza—Destroyer and Teacher," 368.

153 *There was no evidence, of course, that any of these vaccines actually worked:* J. M. Eyler, "The State of Science, Microbiology, and Vaccines Circa 1918," *Public Health Reports* 125, suppl. 3 (2010): 27–36.

154 *Vaccine research didn't kick into overdrive:* Information in the following paragraphs is based on C. Hannoun, "The Evolving History of Influenza Viruses and Influenza Vaccines," *Expert Review of Vaccines* 12, no. 9 (2013): 1085–94.

155 *These centers identify the flu strains:* C. Gerdil, "The Annual Production Cycle for Influenza Vaccine," *Vaccine* 21, no. 16 (2003): 1776–79; "CDC Selecting Viruses for the Flu Season 2016," Centers for Disease Control and Prevention. Accessed April 30, 2018. https://www.cdc.gov/flu/about/season/vaccine-selection.htm.

155 *In the 2004–2005 flu season, that figure was only 10 percent:* Data for the vaccine's efficacy is from "Seasonal Influenza Vaccine Effectiveness, 2005–2018," Centers for Disease Control and Prevention. Accessed May 3, 2018. https://www.cdc.gov/flu/professionals/vaccination/effectiveness-studies.htm.

155 *the 2017–2018 influenza season:* D. McNeil, "It's Not Just You. Lots of People Caught the Flu," *New York Times*, January 19, 2018, A12; D. M. Skowronski et al., "Early Season Co-circulation of Influenza A(H3N2) and B(Yamagata): Interim Estimates of 2017/18 Vaccine Effectiveness, Canada, January 2018," *Eurosurveillance* 23, no. 5 (2018): 18-00035. A more recent paper estimates that overall the vaccine was 36 percent effective, but only 25 percent effective against the H3N2 viruses that were circulating. See B. Flannery et al.,

"Interim Estimates of 2017–18 Seasonal Influenza Vaccine Effectiveness—United States, February 2018," *Morbidity and Mortality Weekly Report* 67 (2018): 180–85.

156 *influenza mortality rate among seniors rose:* W. W. Thompson et al., "Mortality Associated with Influenza and Respiratory Syncytial Virus in the United States," *JAMA* 289, no. 2 (2003): 176–86.

156 *a natural experiment in Japan:* T. A. Reichert et al., "The Japanese Experience with Vaccinating Schoolchildren against Influenza," *New England Journal of Medicine* 344, no. 12 (2001): 889–96.

156 *in contrast to the majority of European countries:* A. McGuire, M. Drummond, and S. Keeping, "Childhood and Adolescent Influenza Vaccination in Europe: A Review of Current Policies and Recommendations for the Future," *Expert Review of Vaccines* 15, no. 5 (2016): 659–70.

156 *Germany provides free vaccines:* Ibid.

156 *almost 60 percent in the U.S:* "Flu Vaccination Coverage, United States, 2015–16 Influenza Season," Centers for Disease Control and Prevention. Accessed April 30, 2018. https://www.cdc.gov/flu/fluvaxview/coverage-1516estimates .htm.

157 *One CDC poster:* Centers for Disease Control and Prevention. Accessed April 30, 2018. https://www.cdc.gov/flu/pdf/freeresources/general/p_universal _question_officeprint.pdf.

157 *recommended the flu vaccine only for those at high risk:* A. E. Fiore et al., "Prevention and Control of Influenza with Vaccines: Recommendations of the Advisory Committee on Immunization Practices (ACIP), 2010," *MMWR Recommendations and Reports* 59, no. RR-8 (2010): 1–62.

158 *It is challenging, therefore, to compare the data:* For the U.S., I used K. D. Kochanek et al., "Deaths: Final Data for 2014," *National Vital Statistics Reports* 65, no. 4 (2016), Table 11. For England, I used NOMIS, run by the government's Office of National Statistics. Available at https://www.nomisweb .co.uk/.

159 *the Cochrane Collaborative in 2014:* V. Demicheli et al., "Vaccines for Preventing Influenza in Healthy Adults," *Cochrane Database of Systematic Reviews* 3 (2014): CD001269.

159 *The CDC describes the flu like this in a poster:* Centers for Disease Control and Prevention. Accessed April 30, 2018. https://www.cdc.gov /immigrantrefugeehealth/pdf/seasonal-flu/flu_and_you_english_508 .pdf. There are a number of CDC informational posters available for physicians to place in their offices.

160 *on the home page of the CDC's flu site:* "About Flu," Centers for Disease Con-

trol and Prevention. Accessed April 30 2018. https://www.cdc.gov/flu/about/
index.html.

160 *advice about the flu from their National Health Service:* "Flu," My Health London. Accessed April 30, 2018. https://www.myhealth.london.nhs.uk/flu.

160 *influenza can be a bit of a nuisance:* Ibid.

161 *Pollard is extremely cognizant of the numerous effects of the flu:* Much of the
information that follows is based on a phone interview with Pollard, July 22,
2017.

162 *In the United States, the cost-effectiveness of the vaccine is less important:*
"Studies in the U.S. have shown that giving the vaccine to healthy working
adults can reduce both the rates of days off work and the number of physician visits. These benefits are seen when the flu vaccine closely matches the
circulating virus. In most years, however, the vaccine does not provide an
overall economic benefit." See C. B. Bridges et al., "Effectiveness and Cost-
Benefit of Influenza Vaccination of Healthy Working Adults: A Randomized
Controlled Trial," *JAMA* 284, no. 13 (2000): 1655–63.

162 *it is recommended for those older than seventy:* "Shingles vaccine FAQs,"
NHS. Accessed April 30, 2018. https://www.nhs.uk/conditions/vaccinations
/shingles-vaccine-questions-and-answers.

162 *"We cannot afford to take a chance":* Sencer and Millar, "Reflections on the
1976 Swine Flu Vaccination Program."

162 *We put more of our patients into the intensive care unit:* S. Murthy and
H. Wunsch, "Clinical Review: International Comparisons in Critical Care—
Lessons Learned," *Critical Care* 16, no. 2 (2012): 218. In the UK there are
about 5 ICU beds per 100,000 people. In the U.S. there are about five times
as many—twenty-five ICU beds per 100,000 people. The UK has a higher life
expectancy than the U.S. See M. Prin and H. Wunsch, "International Comparisons of Intensive Care: Informing Outcomes and Improving Standards,"
Current Opinion in Critical Care 18, no. 6 (2012): 700–706.

163 *We give more chemotherapy to cancer patients:* Ezekiel Emanuel and Justin
Bekelman, "Is It Better to Die in America or in England?," *New York Times,*
January 16, 2016.

10. THE BUSINESS OF FLU

165 *there were an additional 12,000 deaths:* N. Hawkes, "Sharp Spike in Deaths in
England and Wales Needs Investigating, Says Public Health Expert," *British
Medical Journal* 352 (2016): 1981.

165 *in those older than sixty-five, there were 217,000 more deaths:* L. Hiam et al.,

NOTES

"What Caused the Spike in Mortality in England and Wales in January 2015?," *Journal of the Royal Society of Medicine* 110, no. 4 (2017): 131–37.

166 *freed up more than £28 billion in pension liabilities:* "Slowdown in Life Expectancy Could Ease Pension Deficit by £28 Billion, Says Mercer," Mercer. Accessed April 30, 2018. https://www.uk.mercer.com/newsroom/continuous-mortality-investigation-pensions-risk.html.

166 *The Metropolitan Life Insurance Company paid out $24 million:* Crosby, *America's Forgotten Pandemic*, 312.

167 *workers experienced a higher growth in wages:* Thomas Garrett, "War and Pestilence as Labor Market Shocks: U.S. Manufacturing Wage Growth 1914–1919," *Economic Inquiry* 47, no. 4 (2009): 711–25.

167 *The flu's positive impact on per capita income growth:* Elizabeth Brainerd and Mark V. Siegler, "The Economic Effects of the 1918 Influenza Epidemic," CEPR Discussion Paper, no. 3791 (2003).

167 *claiming it had overreached:* These examples are from Thomas Garrett, "Pandemic Economics: The 1918 Infuenza and Its Modern-Day Implications," *Federal Reserve Bank of St. Louis Review* 90, no. 2 (2008): 75–93.

167 *the timetable for banks, theaters, and department stores:* F. Aimone, "The 1918 Influenza Epidemic in New York City: A Review of the Public Health Response," *Public Health Reports* 125, suppl. 3 (2010): 71–79.

167 *the poor were three times more likely to die:* E. Sydenstricker, "The Incidence of Influenza among Persons of Different Economic Status during the Epidemic of 1918," *Public Health Reports* 46, no. 4 (1931): 154–70.

167 *Those who lived in a four-bedroom apartment:* S. E. Mamelund, "A Socially Neutral Disease? Individual Social Class, Household Wealth and Mortality from Spanish Influenza in Two Socially Contrasting Parishes in Kristiania 1918–19," *Social Science & Medicine* 62, no. 4 (2006): 923–40.

167 *Sir Arthur Conan Doyle:* Louise Welsh, "Arthur Conan Doyle's Other Lost World," *Guardian*, May 24, 2009.

167 *Trump's grandfather Frederick:* Gwenda Blair, *The Trumps: Three Generations That Built an Empire* (New York: Touchstone, 2000), 116–17.

168 *Douglas Almond, an economist from Columbia University:* Douglas Almond, "Is the 1918 Influenza Pandemic Over? Long-Term Effects of In Utero Influenza Exposure in the Post-1940 U.S. Population," *Journal of Political Economy* 114, no. 41 (2006): 672–712.

168 *The* Washington Post *was outraged:* "The Ghoulish Coffin Trust," *Washington Post*, October 13, 1918, 4.

168 *took out a full-page ad in the weekly* Moving Picture World: *Moving Picture World*, November 9, 1918, 638.

169 *He died of influenza a week before:* Leslie DeBauche, *Reel Patriotism: The Movies and World War I* (Madison: University of Wisconsin Press, 1997), 149.

169 *Evan Morris was hired by the pharmaceutical giant Roche:* Details about Morris are taken from Brody Mullins, "Hidden Influence: The Fall of K Street's Renegade," *Wall Street Journal*, February 14, 2017, A1; and from an interview Mullins gave to C-SPAN, "Q and A with Brody Mullins," C-SPAN. Accessed April 30, 2018. https://www.c-span.org/video/?424470-1/qa-brody-mullins.

169 *There were 11,000 of them in 2016:* This and the other figures cited are from the Center for Responsive Politics. See "Lobbying Database," OpenSecrets .org. Accessed April 30, 2018. https://www.opensecrets.org/lobby/index.php.

170 *Then he shot himself:* Mullins, "Hidden Influence."

170 *10 percent of the budget of the National Institutes of Health:* This 10 percent set aside began in the 1990s and ended in 2015, when it was no longer required by Congress.

170 *less than 1 percent of the U.S. population is infected with the virus:* "HIV in the United States: At A Glance," Centers for Disease Control and Prevention. Accessed April 30, 2018. https://www.cdc.gov/hiv/statistics/overview /ataglance.html.

171 *Nearly $213 million:* "HHS FY2016 Budget in Brief," U.S Department of Health and Human Services. Accessed April 30, 2018. https://www.hhs.gov /about/budget/budget-in-brief/phssef/index.html.

172 *the budget dropped by 60 percent, to $68 million:* Ibid.

172 *the news was better in 2017:* "HHS FY2017 Budget in Brief," U.S. Department of Health and Human Services. Accessed April 30, 2018. https://www.hhs .gov/about/budget/fy2017/budget-in-brief/phssef/index.html.

172 *The NIH is heavily invested in influenza research:* It is hard to get an exact amount, because reports are generated by looking for any project with a keyword "influenza" or the like, even if the project is really focused on another topic entirely. Even so, it's the best estimate we have. See "Estimates of Funding for Various Research, Condition, and Disease Categories (RCDC)," NIH Portfolio Online Reporting Tools (RePORT). Accessed April 30, 2018. https://report.nih.gov/categorical_spending.aspx.

172 *The NIH estimates that every one dollar it spends on research:* The1:8 ratio is for basic research after eight years. Clinical research yields a more modest 1:2 return on investment after three years. See "Impact of NIH Research," National Institutes of Health. Accessed April 30, 2018. https://www.nih.gov/ about-nih/what-we-do/impact-nih-research/our-society.

172 *in the words of its infomercial:* Cue.Me. Accessed April 30, 2018. https://cue .me/#inflammation.

173 *Cue initially received $2 million from angel investors:* Douglas Macmillan, "Cue Gets $7.5 Million to Build $199 Home Flu-Testing Device," *Wall Street Journal*, November 18, 2014.

173 *"The business is based on the razor-and-blade model":* Ibid.

173 *rapid influenza tests are not very sensitive:* C. Chartrand et al., "Accuracy of Rapid Influenza Diagnostic Tests: A Meta-analysis," *Annals of Internal Medicine* 156, no. 7 (2012): 500–511.

174 *"Total Economic Consequences of an Influenza Outbreak in the United States":* F. Prager, D. Wei, and A. Rose, "Total Economic Consequences of an Influenza Outbreak in the United States," *Risk Analysis* 37, no. 1 (2017): 4–19.

174 *a two-week total electricity blackout in Los Angeles County:* A. Rose, G. Oladosu, and S. Liao, "Business Interruption Impacts of a Terrorist Attack on the Electric Power System of Los Angeles: Customer Resilience to a Total Blackout," *Risk Analysis* 27, no. 3 (2009): 513–31.

175 *the peak activity of the influenza virus came two weeks later:* J. S. Brownstein, C. J. Wolfe, and K. D. Mandl, "Empirical Evidence for the Effect of Airline Travel on Inter-regional Influenza Spread in the United States," *PLoS Medicine* 3, no. 10 (2006): e401.

175 *he heard a news report about a shortage of exotic dancers:* Details are from a phone interview with Charles Stoecker, October 30, 2017.

175 *"Success* Is *Something to Sneeze At":* Charles Stoecker, Nicholas Sanders, and Alan Barreca, "Success *Is* Something to Sneeze At: Influenza Mortality in Regions That Send Teams to the Super Bowl," *Tulane Economics Working Paper Series* 2015; working paper 1501.

176 *headline in the* New York Times: Austin Frakt, "Your Team Made the Super Bowl? Better Get a Flu Shot," *New York Times*, February 1, 2016, A3.

EPILOGUE

180 *Pandemic Influenza Plan:* "Pandemic Influenza Plan, 2017 Update," U.S. Department of Health and Human Services, n.p n.d. Accessed April 30, 2018. https://www.cdc.gov/flu/pandemic-resources/pdf/pan-flu-report-2017v2.pdf.

181 *we still don't have good enough evidence to drive other policy decisions:* T. Jefferson, "Influenza Vaccination: Policy versus Evidence," *British Medical Journal* 333, no. 7574: 912–15.

182 *And so do the viruses we are carrying:* Many years ago, a plane sat on the runway for three hours while a mechanical problem was fixed. It was carrying one passenger with influenza. Three days later, almost three-quarters of the passengers came down with the very same influenza virus. M. R. Moser

et al., "An Outbreak of Influenza aboard a Commercial Airliner," *American Journal of Epidemiology* 110, no. 1 (1979): 1–6.

182 *crowded housing is still a fact of life for many:* "Charting the Progress of Populations," United Nations Population Division. Accessed April 30, 2018. http://www.un.org/esa/population/pubsarchive/chart/14.pdf.

182 *the United States is not immune:* "Historical Census of Housing Tables," The United States Census Bureau. Accessed April 30, 2018. https://www.census .gov/hhes/www/housing/census/historic/crowding.html.

182 *In New York, almost 9 percent of households:* Office of the City Comptroller, "Hidden Households," New York City Housing Brief (2015). Accessed April 30, 2018. https://comptroller.nyc.gov/wp-content/uploads/documents /Hidden_Households.pdf.

183 *A virus that is very potent quickly kills its host:* D. M. Morens, J. K. Taubenberger, and A. S. Fauci, "The Persistent Legacy of the 1918 Influenza Virus," *New England Journal of Medicine* 361, no. 3 (2009): 225–29.

BIBLIOGRAPHY

Abhimanyu and A. K. Coussens. "The Role of UV Radiation and Vitamin D in the Seasonality and Outcomes of Infectious Disease." *Photochemical and Photobiological Sciences* 16, no. 3 (2017): 314–38.

Adams, F. *The Genuine Works of Hippocrates*. New York: William Wood, 1886.

Aimone, F. "The 1918 Influenza Epidemic in New York City: A Review of the Public Health Response." *Public Health Reports* 125, suppl. 3 (2010): 71–79.

Almond, D. "Is the 1918 Influenza Pandemic Over? Long-Term Effects of In Utero Influenza Exposure in the Post-1940 U.S. Population." *Journal of Political Economy* 114, no. 41 (2006): 672–712.

Andrews, C. *The Common Cold*. New York: W. W. Norton, 1965.

Barry, J. M. *The Great Influenza: The Epic Story of the Deadliest Plague in History*. New York: Penguin, 2005.

———. "The Site of Origin of the 1918 Influenza Pandemic and Its Public Health Implications." *Journal of Translational Medicine* 2, no. 1 (2004): 3.

Bergman, P., A. U. Lindh, L. Bjorkhem-Bergman, and J. D. Lindh. "Vitamin D and Respiratory Tract Infections: A Systematic Review and Meta-analysis of Randomized Controlled Trials." *PLoS One* 8, no. 6 (2013): e65835.

Blair, G. *The Trumps: Three Generations That Built an Empire*. New York: Touchstone, 2000.

Boland, M. E., S. M. Roper, and J. A. Henry. "Complications of Quinine Poisoning." *Lancet* 325, no. 8425 (1985): 384–85.

Brett, A. S., and A. Zuger. "The Run on Tamiflu—Should Physicians Prescribe on Demand?" *New England Journal of Medicine* 353, no. 25 (2005): 2636–37.

Bridges, C. B., W. W. Thompson, M. I. Meltzer, G. R. Reeve, W. J. Talamonti, N. J.

Cox, H. A. Lilac, H. Hall, A. Klimov, and K. Fukuda. "Effectiveness and Cost-Benefit of Influenza Vaccination of Healthy Working Adults: A Randomized Controlled Trial." *JAMA* 284, no. 13 (2000): 1655–63.

Brownstein, J. S., C. J. Wolfe, and K. D. Mandl. "Empirical Evidence for the Effect of Airline Travel on Inter-regional Influenza Spread in the United States." *PLoS Medicine* 3, no. 10 (2006): e401.

Butler, D. "When Google Got Flu Wrong." *Nature* 494, no. 7436 (2013): 155–56.

Byerly, C. *Fever of War: The Influenza Epidemic in the U.S. Army during World War I.* New York: New York University Press, 2005.

Cannell, J. J., R. Vieth, J. C. Umhau, M. F. Holick, W. B. Grant, S. Madronich, C. F. Garland, and E. Giovannucci. "Epidemic Influenza and Vitamin D." *Epidemiology & Infection* 134, no. 6 (2006): 1129–40.

Cello, J., A. V. Paul, and E. Wimmer. "Chemical Synthesis of Poliovirus cDNA: Generation of Infectious Virus in the Absence of Natural Template." *Science* 297, no. 5583 (2002): 1016–18.

Centers for Disease Control and Prevention. "Update: Drug Susceptibility of Swine-Origin Influenza A (H1N1) Viruses, April 2009." *Morbidity and Mortality Weekly Report* 58, no. 16 (2009): 433–35.

Chartrand, C., M. M. Leeflang, J. Minion, T. Brewer, and M. Pai. "Accuracy of Rapid Influenza Diagnostic Tests: A Meta-analysis." *Annals of Internal Medicine* 156, no. 7 (2006): 500–11.

Collier, R. *The Plague of the Spanish Lady: The Influenza Pandemic of 1918–1919.* London: Macmillan, 1974.

Cook, S., C. Conrad, A. L. Fowlkes, and M. H. Mohebbi. "Assessing Google Flu Trends Performance in the United States during the 2009 Influenza Virus A (H1N1) Pandemic." *PLoS One* 6, no. 8 (2011): e23610.

Cooper Cole, C. E. "Preliminary Report on Influenza Epidemic at Bramshott in September-October, 1918." *British Medical Journal* 2, no. 3021 (1918): 566–68.

Creighton, C. *A History of Epidemics in Britain.* New York: Barnes & Noble, 1965.

Crosby, A. W. *America's Forgotten Pandemic: The Influenza of 1918.* Cambridge: Cambridge University Press, 2003.

Das, D., K. Mertzger, R. Heffernan, S. Balter, D. Weiss, and F. Mostashari. "Monitoring Over-the-Counter Medication Sales for Early Detection of Disease Outbreaks—New York City." *MMWR Supplements* 54 (2005): 41–46.

Debauche, L. *Reel Patriotism: The Movies and World War I.* Madison: University of Wisconsin Press, 1997.

Demicheli, V., T. Jefferson, L. A. Al-Ansary, E. Ferroni, and C. Di Pietrantonj. "Vaccines for Preventing Influenza in Healthy Adults." *Cochrane Database of Systematic Reviews* 3 (2014): CD001269.

BIBLIOGRAPHY

Dobson, J., R. J. Whitley, S. Pocock, and A. S. Monto. "Oseltamivir Treatment for Influenza in Adults: A Meta-analysis of Randomised Controlled Trials." *Lancet* 385, no. 9979 (2015): 1729–37.

Doshi, P. "The Elusive Definition of Pandemic Influenza." *Bulletin of the World Health Organization* 89 no. 7 (2011): 532–38.

———. "The 2009 Influenza Pandemic." *Lancet Infectious Diseases* 13, no. 3 (2013): 193.

Dow, K., and S. Cutter. "Crying Wolf: Repeat Responses to Hurricane Evacuation Orders." *Coastal Management* 26, no. 4 (1998): 237–52.

Duncan, K. *Hunting the 1918 Flu: One Scientist's Search for a Killer Virus.* Toronto: University of Toronto Press, 2003.

Earn, D. J., P. W. Andrews, and B. M. Bolker. "Population-Level Effects of Suppressing Fever." *Proceedings of the Royal Society B: Biological Sciences* 281, no. 1778 (2014): 20132570.

Edmond, J. D., R. G. Johnston, D. Kidd, H. J. Rylance, and R. G. Sommerville. "The Inhibition of Neuraminidase and Antiviral Action." *British Journal of Pharmacology and Chemotherapy* 27, no. 2 (1966): 415–26.

Emerman, M., and H. S. Malik. "Paleovirology—Modern Consequences of Ancient Viruses." *PLoS Biology* 8, no. 2 (2010): e1000301.

"The Epidemic of Influenza." *JAMA* 71, no. 13 (1918): 1063–64.

Evans, S. S., E. A. Repasky, and D. T. Fisher. "Fever and the Thermal Regulation of Immunity: The Immune System Feels the Heat." *Nature Reviews Immunology* 15, no. 6 (2015): 335–49.

Eyler, J. M. "The State of Science, Microbiology, and Vaccines circa 1918." *Public Health Reports* 125, suppl. 3 (2010): 27–36.

Fabier, F. "Traitement de la Grippe par les Injections de Quinine." *Journal de Médecine et de Chirurgie Pratiques* 90 (1919): 783–84.

Fiore, A. E., T. M. Uyeki, K. Broder, L. Finelli, G. L. Euler, J. A. Singleton, J. K. Iskander et al. "Prevention and Control of Influenza with Vaccines: Recommendations of the Advisory Committee on Immunization Practices (ACIP), 2010." *MMWR Recommendations and Reports* 59, no. RR-8 (2010): 1–62.

Garrett, T. "Pandemic Economics: The 1918 Influenza and Its Modern-Day Implications." *Federal Reserve Bank of St. Louis Review* 90, no. 2 (2008): 75–93.

———. "War and Pestilence as Labor Market Shocks: U.S. Manufacturing Wage Growth 1914–1919." *Economic Inquiry* 47, no. 4 (2009): 711–25.

Gaydos, J. C., F. H. Top Jr., R. A. Hodder, and P. K. Russell. "Swine Influenza A Outbreak, Fort Dix, New Jersey, 1976." *Emerging Infectious Diseases* 12, no. 1 (2006): 23–28.

Gerdil, C. "The Annual Production Cycle for Influenza Vaccine." *Vaccine* 21, no. 16 (2003): 1776–79.

Glatstein, M., and D. Scolnik. "Fever: To Treat or Not to Treat?" *World Journal of Pediatrics* 4, no. 4 (2008): 245–47.

Godlee, F. "Conflicts of Interest and Pandemic Flu." *British Medical Journal* 340 (2010): c2947.

Gregor, A. "A Note on the Epidemiology of Influenza among Workers." *British Medical Journal* 1, no. 3035 (1919): 242–43.

Grijalva, C. G., J. P. Nuorti, and M. R. Griffin. "Antibiotic Prescription Rates for Acute Respiratory Tract Infections in U.S. Ambulatory Settings." *JAMA* 302, no. 7 (2009): 758–66.

Grist, N. R. "Pandemic Influenza 1918." *British Medical Journal* 2, no. 6205 (1979): 1632–33.

Gross, C. P., and K. A. Sepkowitz. "The Myth of the Medical Breakthrough: Smallpox, Vaccination, and Jenner Reconsidered." *International Journal of Infectious Disease* 3, no. 1 (1998): 54–60.

Hammond, G. W., R. L. Raddatz, and D. E. Gelskey. "Impact of Atmospheric Dispersion and Transport of Viral Aerosols on the Epidemiology of Influenza." *Reviews of Infectious Diseases* 11, no. 3 (1989): 494–97.

Hammond, J. A. B., W. Rolland, and T. H. G. Shore. "Purulent Bronchitis: A Study of Cases Occurring amongst the British Troops at a Base in France." *Lancet* 190, no. 4898 (1917): 41–45.

Hannoun, C. "The Evolving History of Influenza Viruses and Influenza Vaccines." *Expert Review of Vaccines* 12, no. 9 (2013): 1085–94.

Hawkes, N. "Sharp Spike in Deaths in England and Wales Needs Investigating, Says Public Health Expert." *British Medical Journal* 352 (2016): i981.

Hayden, F. G., J. J . Treanor, R. F. Betts, M. Lobo, J. D. Esinhart, and E. K. Hussey. "Safety and Efficacy of the Neuraminidase Inhibitor GG167 in Experimental Human Influenza." *JAMA* 275, no. 4 (1996): 295–99.

Henderson, D. A., B. Courtney, T. V. Inglesby, E. Toner, and J. B. Nuzzo. "Public Health and Medical Responses to the 1957–58 Influenza Pandemic." *Biosecurity and Bioterrorism* 7, no. 3 (2009): 265–73.

Herfst, S., E. J. Schrauwen, M. Linster, S. Chutinimitkul, E. De Wit, V. J. Munster, E. M. Sorrell et al. "Airborne Transmission of Influenza A/H5N1 Virus between Ferrets." *Science* 336, no. 6088 (2012): 1534–41.

Hernan, M. A., and M. Lipsitch. "Oseltamivir and Risk of Lower Respiratory Tract Complications in Patients with Flu Symptoms: A Meta-analysis of Eleven Randomized Clinical Trials." *Clinical Infectious Diseases* 53, no. 3 (2011): 277–79.

Herrick, J. B. "Treatment of Influenza by Means Other Than Vaccines and Serums." *JAMA* 73, no. 7 (1919): 482–87.

Hiam, L., D. Dorling, D. Harrison, and M. McKee. "What Caused the Spike in Mortality in England and Wales in January 2015?" *Journal of the Royal Society of Medicine* 110, no. 4 (2017): 131–37.

Hildreth, M. L. "The Influenza Epidemic of 1918–1919 in France: Contemporary Concepts of Aetiology, Therapy, and Prevention." *Social History of Medicine* 4, no. 2 (1991): 277–94.

Hirani, V., and P. Primatesta. "Vitamin D Concentrations among People Aged 65 Years and Over Living in Private Households and Institutions in England: Population Survey." *Age and Ageing* 34 no. 5 (2006): 485–91.

Hirve, S., L. P. Newman, J. Paget, E. Azziz-Baumgartner, J. Fitzner, N. Bhat, K. Vandemaele, and W. Zhang. "Influenza Seasonality in the Tropics and Subtropics—When to Vaccinate?" *PLoS One* 11, no. 4 (2016): e0153003.

Honigsbaum, M. "Regulating the 1918–19 Pandemic: Flu, Stoicism and the Northcliffe Press." *Medical History* 57, no. 2 (2013): 165–85.

Hopkirk, A. F. *Influenza: Its History, Nature, Cause and Treatment.* New York: Walter Scott Publishing, 1914.

Hoyle, F., and C. Wickramasinghe. *Evolution from Space: A Theory of Cosmic Creationism.* New York: Simon & Schuster, 1982.

Hoyle, F., and N. C. Wickramasinghe. "Sunspots and Influenza." *Nature* 343, no. 6256 (1990): 304.

Hughes, S. S. *The Virus: A History of the Concept.* London: Heinemann Educational Books, Science History Publications, 1977.

Influenza Committee of the Advisory Board to the D.G.M.S., France, "The Influenza Epidemic in the British Armies in France, 1918." *British Medical Journal* 2, no. 3019 (1918): 505–9.

"Influenza Discussions." *American Journal of Public Health* 9, no. 2 (1919): 136.

"Influenza: Its History, Nature, Cause and Treatment." Book review. *JAMA* 63, no. 3 (1914): 267.

"Influenza: Kansas—Haskell." *Public Health Reports* 33, no. 14 (1918): 502.

Jack, A. "Flu's Unexpected Bonus." *British Medical Journal* 339 (2009): b3811.

"James B. Herrick (1861–1954)." *JAMA* 16, no. 186 (1963): 722–23.

Jefferson, T. "Influenza Vaccination: Policy versus Evidence." *British Medical Journal* 333, no. 7574 (2006): 912–15.

Jefferson, T., V. Demicheli, D. Rivetti, M. Jones, C. Di Pietrantonj, and A. Rivetti. "Antivirals for Influenza in Healthy Adults: Systematic Review." *Lancet* 367, no. 9507 (2006): 303–13.

BIBLIOGRAPHY

Jefferson, T., and P. Doshi. "Multisystem Failure: The Story of Anti-influenza Drugs." *British Medical Journal* 348 (2014): g2263.

Jefferson, T., M. A. Jones, P. Doshi, C. B. Del Mar, R. Hama, M. J. Thompson, E. A. Spencer et al. "Neuraminidase Inhibitors for Preventing and Treating Influenza in Healthy Adults and Children." *Cochrane Database of Systematic Reviews* 4 (2014): CD008965.

Jenner, E. *An Inquiry into the Causes and Effects of the Variolae Vaccinae, a Disease Discovered in Some of the Western Counties of England, Particularly Gloucestershire, and Known by the Name of the Cow Pox.* London: Sampson Low, 1798.

Kaiser, L., C. Wat, T. Mills, P. Mahoney, P. Ward, and F. Hayden. "Impact of Oseltamivir Treatment on Influenza-Related Lower Respiratory Tract Complications and Hospitalizations." *Archives of Internal Medicine* 163, no. 14 (2003): 1667–72.

Kamat, S., V. Maniaci, M. Y. Linares, and J. M. Lozano. "Pediatric Psychiatric Emergency Department Visits during a Full Moon." *Pediatric Emergency Care* 30, no. 12 (2014): 875–78.

Kelly, H., and K. Grant. "Interim Analysis of Pandemic Influenza (H1N1) 2009 in Australia: Surveillance Trends, Age of Infection and Effectiveness of Seasonal Vaccination." *Eurosurveillance* 14, no. 31 (2009): 1–5.

Kennedy, D. "Better Never Than Late." *Science* 310, no. 5746 (2005): 195.

Khan, A., and W. Patrick. *The Next Pandemic: On the Front Lines against Humankind's Gravest Dangers.* New York: PublicAffairs, 2016.

Kilbourne, E. D. "Influenza Pandemics of the 20th Century." *Emerging Infectious Diseases* 12, no. 1 (2006): 9–14.

Klein, H. A. "The Treatment of 'Spanish Influenza.'" *JAMA* 71, no. 18 (1918): 1510.

Kluger, M. J. *Fever: Its Biology, Evolution, and Function.* Princeton, NJ: Princeton University Press, 1979.

Kmietowicz, Z. "Use Leftover Tamiflu to Grit Icy Roads, MP Suggests." *British Medical Journal* 340 (2010): c501.

Kochanek, K. D., S. L. Murphy, J. Xu, and B. Tejada-Vera. "Deaths: Final Data for 2014." *National Vital Statistics Reports* 65, no. 4 (2016): 1–122.

Kolata, G. *Flu: The Story of the Great Influenza Pandemic of 1918 and the Search for the Virus That Caused It.* New York: Touchstone, 2005.

Lamb, F., and E. Brannin. "The Epidemic Respiratory Infection at Camp Cody N.M." *JAMA* 72, no. 15 (1919): 1056–62.

Langford, C. "Did the 1918–19 Influenza Pandemic Originate in China?" *Population and Development Review* 31, no. 3 (2005): 479–505.

Langmuir, A. D., T. D. Worthen, J. Solomon, C. G. Ray, and E. Petersen. "The Thucydides Syndrome. A New Hypothesis for the Cause of the Plague of Athens." *New England Journal of Medicine* 313, no. 16 (1985): 1027–30.

Lazer, D., R. Kennedy, G. King, and A. Vespignani. "The Parable of Google Flu: Traps in Big Data Analysis." *Science* 343, no. 6176 (2014): 1203–5.

Leary, T. "The Use of the Influenza Vaccine in the Present Epidemic." *American Journal of Public Health* 8, no. 10 (1918): 754–55.

Li, W., S. K. Wong, F. Li, J. H. Kuhn, I. C. Huang, H. Choe, and M. Farzan. "Animal Origins of the Severe Acute Respiratory Syndrome Coronavirus: Insight from ACE2-S-Protein Interactions." *Journal of Virology* 80, no. 9 (2006): 4211–19.

Linder, J. A. "Vitamin D and the Cure for the Common Cold." *JAMA* 308, no. 13 (2012): 1375–76.

Lowen, A. C., S. Mubareka, J. Steel, and P. Palese. "Influenza Virus Transmission Is Dependent on Relative Humidity and Temperature." *PLoS Pathogens* 3, no. 10 (2007): 1470–76.

Malik, M. T., A. Gumel, L. H. Thompson, T. Strome, and S. M. Mahmud. "'Google Flu Trends' and Emergency Department Triage Data Predicted the 2009 Pandemic H1N1 Waves in Manitoba." *Canadian Journal of Public Health* 102, no. 4 (2011): 294–97.

Mamelund, S. E. "A Socially Neutral Disease? Individual Social Class, Household Wealth and Mortality from Spanish Influenza in Two Socially Contrasting Parishes in Kristiania 1918–19." *Social Science & Medicine* 62, no. 4 (2006): 923–40.

McCarthy, M. L., S. L. Zeger, R. Ding, D. Aronsky, N. R. Hoot, and G. D. Kelen. "The Challenge of Predicting Demand for Emergency Department Services." *Academic Emergency Medicine* 15, no. 4 (2008): 337–46.

McGuire, A., M. Drummond, and S. Keeping. "Childhood and Adolescent Influenza Vaccination in Europe: A Review of Current Policies and Recommendations for the Future." *Expert Review of Vaccines* 15, no. 5 (2016): 659–70.

McOscar, J. "Influenza in the Lay Press." *British Medical Journal* 2, no. 3019 (1918): 534.

Mitton, S. *Fred Hoyle: A Life in Science.* Cambridge: Cambridge University Press, 2011.

Molinari, N. A., I. R. Ortega-Sanchez, M. L. Messonnier, W. W. Thompson, P. M. Wortley, E. Weintraub, and C. B. Bridges. "The Annual Impact of Seasonal Influenza in the U.S.: Measuring Disease Burden and Costs." *Vaccine* 25, no. 27 (2007): 5086–96.

Morens, D. M. "Death of a President." *New England Journal of Medicine* 341, no. 24 (1999): 1845–49.

BIBLIOGRAPHY

Morens, D. M., and A. S. Fauci. "The 1918 Influenza Pandemic: Insights for the 21st Century." *Journal of Infectious Disease* 195, no. 7 (2007): 1019–28.

Morens, D. M., G. K. Folkers, and A. S. Fauci. "What Is a Pandemic?" *Journal of Infectious Disease* 200, no. 7 (2009): 1018–21.

Morens, D. M., J. K. Taubenberger, and A. S. Fauci. "The Persistent Legacy of the 1918 Influenza Virus." *New England Journal of Medicine* 361, no. 3 (2009): 225–29.

Moscona, A. "Neuraminidase Inhibitors for Influenza." *New England Journal of Medicine* 353, no. 13 (2005): 1363–73.

Moser, M. R., T. R. Bender, H. S. Margolis, G. R. Noble, A. P. Kendal, and D. G. Ritter. "An Outbreak of Influenza aboard a Commercial Airliner." *American Journal of Epidemiology* 110, no. 1 (1979): 1–6.

Murdoch, D. R., S. Slow, S. T. Chambers, L. C. Jennings, A. W. Stewart, P. C. Priest, C. M. Florkowski, J. H. Livesey, A. C. Camargo, and R. Scragg. "Effect of Vitamin D$_3$ Supplementation on Upper Respiratory Tract Infections in Healthy Adults: The Vidaris Randomized Controlled Trial." *JAMA* 308, no. 13 (2012): 1333–39.

Murray, C. J., A. D. Lopez, B. Chin, D. Feehan, and K. H. Hill. "Estimation of Potential Global Pandemic Influenza Mortality on the Basis of Vital Registry Data from the 1918–20 Pandemic: A Quantitative Analysis." *Lancet* 368, no. 9554 (2006): 2211–18.

Murthy, S., and H. Wunsch. "Clinical Review: International Comparisons in Critical Care—Lessons Learned." *Critical Care* 16, no. 2 (2012): 218.

Nguyen, J. L., J. Schwartz, and D. W. Dockery. "The Relationship between Indoor and Outdoor Temperature, Apparent Temperature, Relative Humidity, and Absolute Humidity." *Indoor Air* 24, no. 1 (2014): 103–12.

Nicolson, J. *The Great Silence, 1918–1920: Living in the Shadow of the Great War.* London: Grove Press, 2010.

Noti, J. D., F. M. Blachere, C. M. McMillen, W. G. Lindsley, M. L. Kashon, D. R. Slaughter, and D. H. Beezhold. "High Humidity Leads to Loss of Infectious Influenza Virus from Simulated Coughs." *PLoS One* 8, no. 2 (2013): e57485.

Olitsky, P., and F. Gates. "Experimental Study of the Nasopharyngeal Secretions from Influenza Patients." *JAMA* 74, no. 22 (1920): 1497–99.

Opie, E., A. Freeman, F. Blake, J. Small, and T. Rivers. "Pneumonia at Camp Funston." *JAMA* 72, no. 2 (1919): 108–13.

Ortiz, J. R., L. Kamimoto, R. E. Aubert, J. Yao, D. K. Shay, J. S. Bresee, and R. S. Epstein. "Oseltamivir Prescribing in Pharmacy-Benefits Database, United States, 2004–2005." *Emerging Infectious Diseases* 14, no. 8 (2008): 1280–83.

Oshinsky, D. M. *Polio: An American Story.* Oxford: Oxford University Press, 2005.

Oxford, J. S. "The So-Called Great Spanish Influenza Pandemic of 1918 May Have Originated in France in 1916." *Philosophical Transactions of the Royal Society of London, Series B: Biological Sciences* 356, no. 1416 (2001): 1857–59.

Patterson, K. D. *Pandemic Influenza, 1700–1900.* Totowa, NJ: Rowman and Littlefield, 1986.

Patwardhan, A., and R. Bilkovski. "Comparison: Flu Prescription Sales Data from a Retail Pharmacy in the U.S. with Google Flu Trends and U.S. ILINet (CDC) Data as Flu Activity Indicator." *PLoS One* 7, no. 8 (2012): e43611.

Petersen, W. F., and S. A. Levinson. "The Therapeutic Effect of Venesection with Reference to Lobar Pneumonia." *JAMA* 78, no. 4 (1922): 257–58.

Polgreen, P. M., F. D. Nelson, and G. R. Neumann. "Use of Prediction Markets to Forecast Infectious Disease Activity." *Clinical Infectious Diseases* 44, no. 2 (2007): 272–79.

Prager, F., D. Wei, and A. Rose. "Total Economic Consequences of an Influenza Outbreak in the United States." *Risk Analysis* 37, no. 1 (2017): 4–19.

Price, G. M. "Influenza—Destroyer and Teacher." *Survey* 41, no. 12 (1918): 367–69.

"Proceedings of the Forty-Sixth Annual Meeting of the American Public Health Association." *American Journal of Public Health* 9, no. 2 (1919): 130–42.

Reichert, T. A., N. Sugaya, D. S. Fedson, W. P. Glezen, L. Simonsen, and M. Tashiro. "The Japanese Experience with Vaccinating Schoolchildren against Influenza." *New England Journal of Medicine* 344, no. 12 (2001): 889–96.

Reid, A. H., T. A. Janczewski, R. M. Lourens, A. J. Elliot, R. S. Daniels, C. L. Berry, J. S. Oxford, and J. K. Taubenberger. "1918 Influenza Pandemic Caused by Highly Conserved Viruses with Two Receptor-Binding Variants." *Emerging Infectious Diseases* 9, no. 10 (2003): 1249–53.

Rice, E. W., N. J. Adcock, M. Sivaganesan, J. D. Brown, D. E. Stallknecht, and D. E. Swayne. "Chlorine Inactivation of Highly Pathogenic Avian Influenza Virus (H5N1)." *Emerging Infectious Diseases* 13, no. 10 (2007): 1568–70.

Riedel, S. "Edward Jenner and the History of Smallpox and Vaccination." *Proceedings (Baylor University Medical Center)* 18, no. 1 (2005): 21–25.

Robins, N. S. *Copeland's Cure: Homeopathy and the War between Conventional and Alternative Medicine.* New York: Knopf, 2005.

Rose, A., G. Oladosu, and S. Liao. "Business Interruption Impacts of a Terrorist Attack on the Electric Power System of Los Angeles: Customer Resilience to a Total Blackout." *Risk Analysis* 27, no. 3 (2009): 513–31.

Rosenow, E. "Prophylactic Inoculation against Respiratory Infections." *JAMA* 72, no. 1 (1919): 31–34.

BIBLIOGRAPHY

Ross, R. S. "A Parlous State of Storm and Stress. The Life and Times of James B. Herrick." *Circulation* 67, no. 5 (1983): 955–59.

Saketkhoo, K., A. Januszkiewicz, and M. A. Sackner. "Effects of Drinking Hot Water, Cold Water, and Chicken Soup on Nasal Mucus Velocity and Nasal Airflow Resistance." *Chest* 74, no. 4 (1978): 408–10.

Saunders-Hastings, P. R., and D. Krewski. "Reviewing the History of Pandemic Influenza: Understanding Patterns of Emergence and Transmission." *Pathogens* 5, no. 4 (2016): 66.

Sencer, D. J., and J. D. Millar. "Reflections on the 1976 Swine Flu Vaccination Program." *Emerging Infectious Diseases* 12, no. 1 (2006): 29–33.

Shadrin, A. S., I. G. Marinich, and L. Y. Taros. "Experimental and Epidemiological Estimation of Seasonal and Climato-Geographical Features of Non-Specific Resistance of the Organism to Influenza." *Journal of Hygiene, Epidemiology, Microbiology, and Immunology* 21, no. 2 (1977): 155–61.

Shaman, J., and A. Karspeck. "Forecasting Seasonal Outbreaks of Influenza." *Proceedings of the National Academy of Sciences* 109, no. 50 (2012): 20425–30.

Shaman, J., A. Karspeck, W. Yang, J. Tamerius, and M. Lipsitch. "Real-Time Influenza Forecasts during the 2012–2013 Season." *Nature Communications* 4 (2013): 2837.

Shanks, G. D., S. Burroughs, J. D. Sohn, N. C. Waters, V. F. Smith, M. Waller, and J. F. Brundage. "Variable Mortality from the 1918–1919 Influenza Pandemic during Military Training." *Military Medicine* 181, no. 8 (2016): 878–82.

Sherertz, R. J., and H. J. Sherertz. "Influenza in the Preantibiotic Era." *Infectious Diseases in Clinical Practice* 14, no. 3 (2006): 127.

Shortridge, K. F. "The 1918 'Spanish' Flu: Pearls from Swine?" *Nature Medicine* 5, no. 4 (1999): 384–85.

Shrestha, S. S., D. L. Swerdlow, R. H. Borse, V. S. Prabhu, L. Finelli, C. Y. Atkins, K. Owusu-Edusei et al. "Estimating the Burden of 2009 Pandemic Influenza A (H1N1) in the United States (April 2009–April 2010)." *Clinical Infectious Diseases* 52, suppl. 1 (2011): S75–82.

Shufflebotham, F. "Influenza among Poison Gas Workers." *British Medical Journal* 1, no. 3042 (1919): 478–79.

Skowronski, D. M., C. Chambers, G. De Serres, J. A. Dickinson, A. L. Winter, R. Hickman, T. Chan et al. "Early Season Co-circulation of Influenza A(H3N2) and B(Yamagata): Interim Estimates of 2017/18 Vaccine Effectiveness, Canada, January 2018." *Eurosurveillance* 23, no. 5 (2018): DOI: 10.3201/eid1201.051254.

Smith, D. C. "Quinine and Fever: The Development of the Effective Dosage."

Journal of the History of Medicine and Allied Sciences 31, no. 3 (1976): 343–67.

Smith, D. W., and B. S. Bradshaw. "Variation in Life Expectancy during the Twentieth Century in the United States." *Demography* 43, no. 4 (2006): 647–57.

Smith, W., C. Andrewes, and P. Laidlaw. "A Virus Obtained from Influenza Patients." *Lancet* 2, no. 5723 (1933): 66–68.

Sneader, W. *Drug Discovery: A History.* Hoboken, NJ: Wiley & Sons, 2005.

Sorbello, A., S. C. Jones, W. Carter, K. Struble, R. Boucher, M. Truffa, D. Birnkrant et al. "Emergency Use Authorization for Intravenous Peramivir: Evaluation of Safety in the Treatment of Hospitalized Patients Infected with 2009 H1N1 Influenza A Virus." *Clinical Infectious Diseases* 55, no. 1 (2012): 1–7.

Spurgeon, D. "Roche Canada Stops Distributing Oseltamivir." *British Medical Journal* 331, no. 7524 (2005): 1041.

Starko, K. M. "Salicylates and Pandemic Influenza Mortality, 1918–1919: Pharmacology, Pathology, and Historic Evidence." *Clinical Infectious Diseases* 49, no. 9 (2009): 1405–10.

Stern, H. *Theory and Practice of Bloodletting.* New York: Rebman Company, 1915.

Stoecker, C., N. Sanders, and A. Barreca. "Success *Is* Something to Sneeze At: Influenza Mortality in Regions That Send Teams to the Super Bowl." *Tulane Economics Working Paper Series* 2015; working paper 1501.

Sydenstricker, E. "The Incidence of Influenza among Persons of Different Economic Status during the Epidemic of 1918." *Public Health Reports* 46, no. 4 (1931): 154–70.

Taubenberger, J. K. "The Origin and Virulence of the 1918 'Spanish' Influenza Virus." *Proceedings of the American Philosophical Society* 150, no. 1 (2006): 86–112.

Taubenberger, J. K., J. V. Hultin, and D. M. Morens. "Discovery and Characterization of the 1918 Pandemic Influenza Virus in Historical Context." *Antiviral Therapy* 12, no. 4, part B (2007): 581–91.

Taubenberger, J. K., A. H. Reid, and T. G. Fanning. "Capturing a Killer Flu Virus." *Scientific American* 292, no. 1 (January 2005): 62–71.

Taubenberger, J. K., A. H. Reid, A. E. Krafft, K. E. Bijwaard, and T. G. Fanning. "Initial Genetic Characterization of the 1918 'Spanish' Influenza Virus." *Science* 275, no. 5307 (1997): 1793–96.

Taubenberger, J. K., A. H. Reid, R. M. Lourens, R. Wang, G. Jin, and T. G. Fanning. "Characterization of the 1918 Influenza Virus Polymerase Genes." *Nature* 437, no. 7060 (2005): 889–93.

Thompson, W. W., D. K. Shay, E. Weintraub, L. Brammer, N. Cox, L. J. Anderson,

BIBLIOGRAPHY

and K. Fukuda. "Mortality Associated with Influenza and Respiratory Syncytial Virus in the United States." *JAMA* 289, no. 2 (2003): 179–86.

Tiwari, Y., S. Goel, and A. Singh. "Arrival Time Pattern and Waiting Time Distribution of Patients in the Emergency Outpatient Department of a Tertiary Level Health Care Institution of North India." *Journal of Emergencies, Trauma, and Shock* 7, no. 3 (2014): 160–65.

Tumpey, T. M., C. F. Basler, P. V. Aguilar, H. Zeng, A. Solorzano, D. E. Swayne, N. J. Cox et al. "Characterization of the Reconstructed 1918 Spanish Influenza Pandemic Virus." *Science* 310, no. 5745 (2005): 77–80.

"Undetermined Disease—Valencia." *Public Health Reports* 33, no. 26 (1918): 1087.

Urashima, M., T. Segawa, M. Okazaki, M. Kurihara, Y. Wada, and H. Ida. "Randomized Trial of Vitamin D Supplementation to Prevent Seasonal Influenza A in Schoolchildren." *American Journal of Clinical Nutrition* 91, no. 5 (2010): 1255–60.

Valdivia, A., J. Lopez-Alcalde, M. Vicente, M. Pichiule, M. Ruiz, and M. Ordobas. "Monitoring Influenza Activity in Europe with Google Flu Trends: Comparison with the Findings of Sentinel Physician Networks—Results for 2009–10." *Eurosurveillance* 15, no. 29 (2010): ii:19621.

Vaughan, V. C. *Doctor's Memories.* New York: Bobbs-Merrill Company, 1926.

Von Alvensleben, A. "Influenza According to Hoyle." *Nature* 344, no. 6265 (1990): 374.

Watanabe, T., G. Zhong, C. A. Russell, N. Nakajima, M. Hatta, A. Hanson, R. McBride, et al. "Circulating Avian Influenza Viruses Closely Related to the 1918 Virus Have Pandemic Potential." *Cell Host & Microbe* 15, no. 6 (2014): 692–705.

Welch, S. J., S. S. Jones, and T. Allen. "Mapping the 24-Hour Emergency Department Cycle to Improve Patient Flow." *Joint Commission Journal on Quality and Patient Safety* 33, no. 5 (2007): 247–55.

Welsch, R. *A Treasury of Nebraska Pioneer Folklore.* Lincoln: University of Nebraska Press, 1967.

Winternitz, M. C., I. M. Wason, and F. P. McNamara. *The Pathology of Influenza.* New Haven, CT: Yale University Press, 1920.

Wootton, D. *Bad Medicine: Doctors Doing Harm since Hippocrates.* Oxford: Oxford University Press, 2006.

Yamanouchi, T., K. Sakakami, and S. Iwashima. "The Infecting Agent in Influenza: An Experimental Research." *Lancet* 193, no. 4997 (1919): 971.

Yang, Y., A. V. Diez Roux, and C. R. Bingham. "Variability and Seasonality of Active Transportation in USA: Evidence from the 2001 NHTS." *International Journal of Behavioral Nutrition and Physical Activity* 8 (2011): 96.

BIBLIOGRAPHY

Zadshir, A., N. Tareen, D. Pan, K. Norris, and D. Martins. "The Prevalence of Hypovitaminosis D among U.S. Adults: Data from the NHANES III." *Ethnicity & Disease* 15, no. 4, suppl. 5 (2005): S5–97–101.

Zhang, X., M. I. Meltzer, and P. M. Wortley. "FluSurge—a Tool to Estimate Demand for Hospital Services during the Next Pandemic Influenza." *Medical Decision Making* 26, no. 6 (2006): 617–23.

INDEX

INDEX

British Medical Journal, 18, 49, 57, 79,
  144, 146
bubonic plague, 62
Burchess, Mark, 132, 133, 134
Bush, George W., 137–38, 139, 170
business: effects of health care on,
  165–77

C virus strain, 37
California: and collecting/reporting
  data, 101
calomel, 13
camel milk, 6
Camp Devens (Massachusetts), 49–51,
  153
Camp Dodge (Iowa), 51
Camp Funston (Kansas), 44, 47, 51
Camp Upton (New York State), 51
camphorated oil, 16
Canada, 69, 139, 145
carbon monoxide, 23
castor oil, 16, 27
CDC. *See* Centers for Disease Control
cells, 36, 39–40. *See also type of cell*
Centers for Disease Control (CDC)
  antiviral medications guidelines of,
    145
  and collecting/reporting data, 101,
    102, 103, 106, 107–9, 114, 115,
    116, 117
  and forecasting the flu, 103
  and Fort Dix outbreak, 70, 71, 72
  funding for, 171
  and GBS cases, 74
  Gupta visit to, 113
  and Hong Kong outbreak (1968), 80
  humidity theory and, 126
  and Nelson studies of health care
    workers, 107–8
  and Ropes & Gray Tamiflu
    prescriptions, 140
  and samples of 1918 virus strain, 90,
    93, 96
  Shaman forecasting model and, 128
  and stockpile of emergency
    medicines, 131, 133, 134, 147

  Tamiflu controversy and, 144,
    145–46, 147
  and 2009 pandemic, 77, 79, 140
  and vaccines/vaccinations, 74, 75,
    155, 156, 157, 159–60, 162
  See also National Center for
    Immunization and Respiratory
    Diseases; Strategic National
    Stockpile
champagne, 15, 27
chemotherapy, 22
chest X-rays, 2, 25. *See also* lung X-rays
Chicago, Illinois: public health officials
  meeting (1918) in, 61–63, 76
chicken cholera, 152
chicken pox, 37, 40, 162
chicken soup, 9–10, 21
chickens
  and cultivation of flu virus, 65
  Egyptian outbreak and, 115–16, 117
  and 1968 pandemic, 69
  and vaccine development, 69, 154
  See also avian flu
children
  and collecting/reporting data, 102
  and complications of influenza, 160
  1957 pandemic and, 68
  and syncytial virus, 101–2
  Tamiflu and, 135–36, 137, 138
  2009 pandemic and, 76
  and vaccines, 155, 156, 158, 162, 163
chills, 9, 21, 31, 99, 101
China, 44–45, 48, 116, 182
chlorine, 18
"circling the drain," 3
climate
  relationship between influenza and, 89
  See also humidity; seasonality;
    weather
Clinton, Bill, 82
Cochrane Collaborative, 137, 139–45,
  146, 147, 159
codeine, 16
cold water: as medical treatment, 9
colds, 9, 101
collective memory, 184–85

243